BURNING PLANET

THE STORY OF FIRE THROUGH TIME

ANDREW C. SCOTT

OXFORD
UNIVERSITY PRESS

OXFORD

UNIVERSITY PRESS

Great Clarendon Street, Oxford, OX2 6DP,
United Kingdom

Oxford University Press is a department of the University of Oxford.
It furthers the University's objective of excellence in research, scholarship,
and education by publishing worldwide. Oxford is a registered trade mark of
Oxford University Press in the UK and in certain other countries

Published in the United States of America by Oxford University Press
198 Madison Avenue, New York, NY 10016, United States of America

British Library Cataloguing in Publication Data
Data available

Library of Congress Control Number: 2017947208

ISBN 978-0-19-873484-0

Printed in Great Britain by
Clays Ltd, St Ives plc

Links to third party websites are provided by Oxford in good faith and
for information only. Oxford disclaims any responsibility for the materials
contained in any third party website referenced in this work.

For Bill Chaloner FRS (1928–2016),
for introducing me to the world of fire in deep time,
and for my wife Anne, for her patience

CONTENTS

PREFACE

There is something exhilarating as well as terrifying about a wildfire. Accounts and images of wildfires raging out of control form dramatic news stories. In this context, fire is invariably regarded as bad, and assumed to have been started by deliberate acts of arson. Yet this is not always the case. We tend to forget that fire is also a force of nature. Fire on Earth today is much misunderstood, and few are aware that it has a history on our planet that stretches back 400 million years. Evidence of fire has now been found deep in the Earth's past. In this book, I describe what has been discovered of the long history of fire, and our understanding of the role that fire has played in evolution and ecology, as well as its taming by humans, and the challenge that wildfire poses today, and is likely to pose increasingly as the Earth warms.

I have spent my working life researching fire. My paper on 300-million-year-old charcoalified leaves—the first paper identifying the earliest fossil conifer—was published more than 40 years ago. Since then, I have worked on preservation by fire and what it can tell us about Earth history. The incredible preservation of charcoal produced from wildfire can be a revelation. It can capture the finest details of flowers and other delicate plant organs. Through the rich information provided by fossil charcoal, we can piece together the long history of fire on this planet, the vegetation it burned, and the climate in which that vegetation grew.

I hope this book will introduce fire and the remarkable story that fossil charcoal holds to anyone interested in the workings and the history of our planet. Occasionally I will have to introduce geological terms that may be new to some readers, though they are few, and explained on their first mention. But to help those who are less familiar with the language of geology, there is a short glossary of terms that appear from time to time through the book.

When we talk about deep history, we will need to use the International Geological Time Scale, which divides the millions of years of Earth history into eras, periods, and epochs. It is reproduced at the back of the book for quick reference (see Appendix).

I must first thank the late Professor William G. (Bill) Chaloner FRS, who first awoke my interest in the deep-time history of wildfire while I was his PhD student. It was a pleasure to have shared an office with him when we were both retired from our academic teaching positions. I must also thank Margaret Collinson (now Professor Margaret E. Collinson), a fellow student of Bill's (we started together), who has also been my colleague at Royal Holloway for more than 20 years, and who has shared my interest in charcoal and wildfire; and Ian Glasspool, my former research student, who has been a great encouragement and help to me over the past 20 years. Many other former research students of ours, including Mick Cope, Kate Bartram, Richard Bateman, Tim Jones, Rachel Brown, Howard Falcon-Lang, Claire Belcher, Laura McParland, Vicky Hudspith, Mark Hardiman, Sarah Brown, and Brittany Robson, have all been of help in shaping my ideas. I also thank my colleagues at Royal Holloway, Gary Nichols, Dave Mattey, Dave Waltham, Sharon Gibbons, Neil Holloway, Kevin d'Souza, as well as research assistants and post-doctoral fellows Nick Rowe, Jenny Cripps, and David Steart for encouragement.

My experience of modern fires was developed through contact firstly with Deborah Martin, John Moody, and Susan Canon, and then through the Pyrogeography Research Group that was put together by David Bowman and Jennifer Balch. I thank them for inviting me to participate. I thank the Department of Geology and Geophysics at Yale University for the award of a visiting professorship that allowed me time to shape my ideas, and the late Bob Berner, the late Leo Hickey, the late Karl Turekian, and Derek Briggs for making this possible. Thanks also are due to Stephen Pyne, William Bond, Chris Roos, and many others who have encouraged the development of my ideas. My good friend Justin Champion (Professor of the History of Modern Ideas at Royal Holloway, University of London) has helped me with some historical aspects of this book, and his encouragement is acknowledged.

I thank Steve Greb, Ian Glasspool, Gary Nichols, Stefan Doerr, Tom Swetnam, Deborah Martin, John Moody, Margaret Collinson, Stuart Baldwin, Dan Neary, Douglas Henderson, Min Minnie Wong, Pat Bartlein, Jenny Marlon, Sally Archibald, John Gowlett, LeRoy Westerling, and Guido van der Werf for kindly supplying illustrations for the book.

I particularly thank my editor, Latha Menon, for inviting me to write this book, and without whom it would not have taken shape; and assistant editor Jenny Nugee of OUP for seeing me through the publication process. I also thank Dan Harding for copy-editing and Gemma Wilkins my production editor. I thank the two official readers of the book and Richard Wright, who made useful suggestions.

Finally, I thank my wife, Anne, and children, Rob and Katrina, for their patience and encouragement over more than 30 years.

1

Introducing fire

Gie me a spark o' Nature's fire,
That's a' the learning I desire.

Robert Burns, 1786

Fire has a bad reputation. Wildfires raging across parts of California and Australia make headlines. In the news bulletins, it is a destructive force that has to be quenched. But that is far from the whole story. Fire has a long history. In our deep past, wildfire helped shape aspects of our planet, and plants and animals have adapted to it in a variety of ways. In this book, we will follow the story of fire through time. But we begin with the present, with the fires that occur around the world today, and how satellites are changing our view of wildfire.

Most of us have little or no experience of a wildfire, apart from those dramatic scenes shown on our television sets from time to time. Almost invariably, two questions are asked: who started the fire, and how quickly can it be put out? Reasonable though they seem, these two questions betray a potential misunderstanding of how fire works on our planet. We assume that the fire was started by humans, either accidentally or deliberately. This may indeed be true, but more than half of the fires started across the globe have a natural cause—mostly lightning strikes, but also other causes such as volcanic activity. Every moment of every day, a fire is burning somewhere in the world. The second assumption is that a fire should always be suppressed. But should we always be rushing to put out a vegetation fire?

Wildfire is one of nature's most frightening manifestations. Winds and storms may die down, and we can seek shelter from them, but fire can be difficult to outrun and escape. Many who are killed by wildfire have underestimated this force of nature, and even those with experience in putting out fires can find themselves cut off, and succumbing to the flames.

As we shall discover, not all vegetation burns in the same way, and there are many different kinds of fire, from those burning surface vegetation to those moving through the crowns of trees. Their consequences may also be very different. In some parts of the world, fire is not only a natural phenomenon but an essential element of the ecosystem. In other environments, fire is unnatural and to be avoided. This may seem a simple dichotomy, but one region, or more especially country, may encompass both situations, so that a national policy concerning fire is not only difficult to formulate but also very difficult to implement. In the case of Madagascar, for example, one half of the island needs fire and the other half does not, so a single country-wide fire suppression policy has caused many unintended consequences to the environment.[1] In some circumstances, putting out a fire may lead eventually to an even more intense and severe fire that could be much more damaging. We are left with a mosaic of 'good' and 'bad' fires, though all tend to be branded as bad.

Southern England, where I currently live, is not an area which experiences many fires. When they do occur, they are characterized as disastrous. I remember a fire near my house, in heather heathland, following which it looked as if everything had been destroyed. The fire left a blackened, desolate landscape, and the news reports carried bleak images and the suggestion that this

(a)

(b)

Figure 1. (a) The aftermath of a fire at Frensham in Surrey, England, in 1995, showing a black landscape and the growth of ferns soon after; (b) the same area ten years later, showing regrowth of heather.

was a calamity. Yet visiting the area today, more than 20 years later, there is no evidence at all that there was a fire—the vegetation has completely recovered (Figure 1). We need a more informed view of wildfire, and that means developing a much greater understanding of the role of fire as part of the Earth system.

To appreciate the role of fire on Earth, we need to look at what the phenomenon of fire involves: what elements create a fire; what is needed to sustain a fire; when is a fire advantageous for an ecosystem, and when is it not? To assess fire in terms of the Earth system, we must consider not only the physical and chemical elements of fire, but also the ecological and environmental aspects. Human involvement in fire is, geologically speaking, relatively recent, but our impact on the landscape has been considerable, and must also be borne in mind.

Fire has been an important part of how the Earth works for at least 400 million years, ever since there has been enough vegetation on land to support a fire. Fires need fuel. They also need oxygen, which is, after all, what allows a fire to burn: combustion is a combination of another element with oxygen, a chemical reaction that gives out heat and light. So there must be enough oxygen in the atmosphere, and that took time to build up in Earth history. And finally, the climate matters. The fuel needs to be dry for a fire to take hold. This is more often the case in warmer climates, but prolonged periods of low rainfall can create conditions for a fire to start and spread. Wind may also be a factor in the rapid spread of a fire. This has led to the concept of fire weather and even of 'fire seasons', periods of time during which fires may be expected. Our increasing ability to predict the occurrence of fire has significantly reduced fatalities in wildfires.

Even in the most promising conditions, fires cannot start without a source of ignition, which could be natural or human, accidental or deliberate. The tendency has been to look for someone to blame—from the deliberate arsonist to someone acting carelessly with a campfire, barbeque, or cigarette. In all this 'blame game' we tend to forget that some vegetation is more likely to burn than others, and that starting a fire in some regions is easier than in others, and there may be unexpected or unintended consequences. I remember not so long ago watching a television documentary about fires in California.[2] The presenter was telling the audience how some of the vegetation was easily set alight and fanned by strong Santa Anna winds, and he wanted to show that people were building their communities in some very flammable vegetation. To make the point he drove up a hill to film the vista, and stopped his vehicle on the grass verge. However, he had a car with a catalytic converter (supposedly environmentally friendly) and did not appreciate that the underneath of the chassis would be hot. The grass under the vehicle caught fire, the car exploded, and a wildfire began. It was fanned by the wind, spread to the dry surrounding vegetation, and moved towards the houses. The rest of the film was taken up with showing how the fire took hold, and how firefighters fought to save the nearby community.

Tracking modern fires

Our understanding of the role of fire on the modern Earth has only really come in the past 30 years or so, and the development of real-time satellite imaging was the key. Combined with the rapid advance in technology that enables pictures from remote places to reach our living rooms, and the advent of the Internet, we are more aware than ever of the force of nature that we call a wildfire. Yet public understanding of wildfire has not kept pace.

For those living in rural communities, fire has sometimes been seen as a positive feature, under human control. Slash-and-burn agriculture may be followed by a more sustainable use of fire to clear fields or to change the use of agricultural land. As more and more people have moved into urban centres though, fire, at least wildfire, has been excluded. So we may ask, where does fire occur on Earth? Where is it a natural part of the ecosystem, and where does it not belong?

I myself had little understanding of the vast areas of vegetation on Earth that burn on a regular basis, or of how humans have interacted with nature to limit, control, or even, in some cases, to promote fire. It is only when you see large tracts of land from the air or even from space that the scale of some fires can be appreciated.

Until the late 1960s, the size and impact of a fire had to be based on rather spotty observations and assessments, such as airport visibility. The first breakthrough arose not from air surveying but satellite imagery. In the 1970s, imagery became available from Landsat satellites, the first of which was launched in 1972. These satellites systematically photograph the Earth's surface, allowing fires to be tracked by comparing daily images of a burning area. Moreover, the full size of a burned area could be mapped. Another significant breakthrough resulted from the ability to record and analyse different parts of the light spectrum. The use of infrared spectroscopy, in particular, enabled living vegetation, which shows up red in the photographs, to be distinguished from vegetation killed by fire, and indeed the absence of vegetation, all of which could be accurately mapped and quantified. This feature of data acquisition from a Landsat image allowed the development of false colour images that played a pivotal role, for example, in the development in the USA of Burned Area Emergency Response burn severity maps, which allowed planners and foresters to consider the aftermath of a fire and the follow-up measures needed. Landsat imagery gave us a more regional understanding of landscape-wide fires.

It was not until the 1980s, however, that further developments came with the use of new satellite data. This period saw the development of the Advanced Very High Resolution Radiometer, an instrument carried by satellite that can be used to scan the Earth's surface at various long-wavelength bands. With continuous upgrading, it can now capture a wide variety of data by measuring different wavelengths. Smoke plumes from fires can be identified, and the temperatures of fires derived from thermal infrared data. One of the key developments was to be able to monitor active fires during night-time.

Satellites can be placed in a variety or orbits. Some, known as geostationary satellites, remain above the same spot on the Earth, so they can be set up to record data at that spot continuously as the Earth spins on its axis. By contrast, polar-orbiting satellites can provide whole-world coverage over a 24-hour period. All these satellites allow observations on a daily basis.

Today we have many satellites, launched by NASA, the European Space Agency (ESA), and various individual countries, which together provide a wide range of fire-relevant data. ESA's ENVISAT, for instance, was devised to monitor climate and environmental changes. This satellite has a polar orbit and carries the AATSR (Advanced Along Track Scanning Radiometer), which has provided data used to construct a fire atlas, giving the precise locations of fires, by processing more than 80,000 images each year.[3] One of the most useful instruments on NASA's polar-orbiting Terra and Aqua satellites is MODIS (Moderate Resolution Imaging Spectroradiometer). Data from this sensor is used to map active fires and to calculate the size of areas burned. The amount of data produced by this sensor has been astonishing. Add to this the vast quantities of data from an array of satellites with increasingly sophisticated instruments and it is evident that the breakthroughs in understanding, following, and predicting fires could not have been achieved without parallel developments in computing that have enabled this 'big data' to be managed, organized, and interpreted.

Some of the most dramatic fire images have come from the International Space Station. The peat fires of Indonesia can be seen burning from many dozens of centres, and their smoke plumes merging together as they drift northwards towards Singapore and Malaysia. A single image from southern California, from October 2007, shows many fires with the smoke plumes billowing over the Pacific Ocean (Colour Plate 1). It is only when we see images such as these from space that we realize the scope and scale of wildfire on Earth. Its scale is even more striking when these data are displayed on a summative year map showing the world distribution of wildfire (Colour Plate 2). After all, there are many fires burning around the world every second of every day.

Data on fire occurrence can show how fires have changed in their distribution on a monthly cycle. Nowhere is this more dramatic than Africa; through the year we can see the areas being burned move southwards (Colour Plate 3). Another surprise has been the ability to see political boundaries where there are different approaches to fire management in two different countries. A good example is the occurrence of fire on either side of the Russian–Chinese border, where the boundary in the Far East can be picked out by fires occurring on the Russian side of the border (Figure 2).

All these developments have led to advanced computer models such as FARSITE, used by US government agencies. FARSITE computes wildfire growth and behaviour for long time periods under different conditions of terrain, fuels, and weather. Such models are providing insights into how fire will be modified in the future under a changing climate.[4]

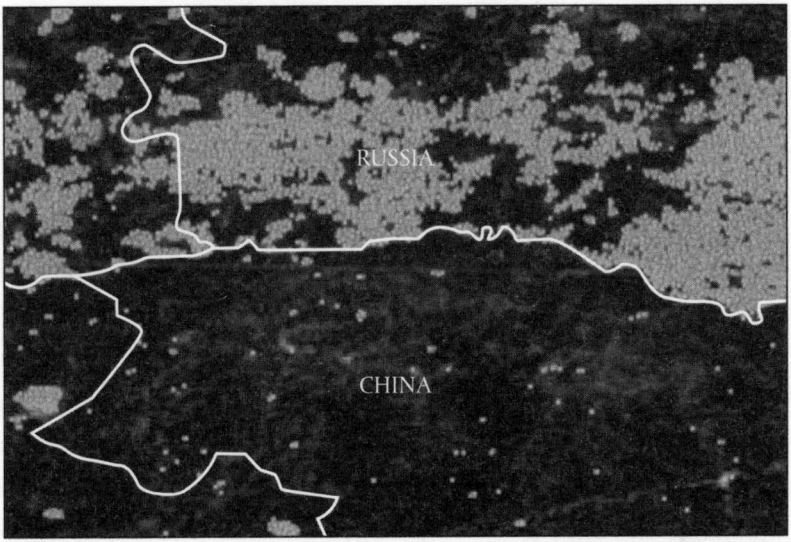

Figure 2. Fires mapped from space. Fires are shown by the pale dots. The line of fire clearly shows the boundary between Russia and China.

After the fire

We are all guilty of not thinking beyond the headlines. We may see a major fire reported on television or in the newspapers but don't think of what happens after. The aftermath is not confined to the rebuilding of communities and regrowth of vegetation.

I was as guilty as anyone in not appreciating the consequences of a major wildfire. My learning experience came when I visited Denver, Colorado, in October 2002. Earlier that year there had been a major fire called the 'Hayman Fire', which covered a vast area outside Denver, surrounding the main water reservoir of the city. That year Denver was the site of the Geological Society of America's annual convention, and I used the opportunity to go out and see the area of the fire with researchers from the United States Geological Survey. It was the first time I had seen the aftermath of a major forest fire. The snow had come early that year, but even so I was to get quite a surprise.

When I had seen on the news vast areas of forest being ravaged by fire, I had always assumed that the trees would have been completely consumed by the fire. But even in areas that saw some of the fiercest burning, most of the tree trunks were still standing, even though the trunks of many were blackened and all the leaves combusted (Figure 3). Some trees and bushes had not been completely burned, and some still had their leaves; a few nearer the forest floor were only browned and still living. But the ground cover was destroyed, either completely or converted to charcoal. It was also evident that not all trees and areas had been affected in the same way.

Figure 3. Stands of pine trees after the 2002 Hayman Fire near Denver, Colorado, showing charred and un-charred trees.

The second feature that I had not expected to see was the movement, by water, of charcoal that had been produced by the fire, together with sediment from the soil layer. This is called *post-fire erosion*. The burning of surface vegetation has two main effects. First, it kills the plants and burns up or chars the plant material. Second, the heat may affect the soil and its structure. Organic material may be destroyed in one part of the soil but some organic compounds may also be deposited in other parts, creating a water-repellent layer within the soil, beneath the surface. The impact of this may be profound. The loss of the plant root systems that bind the soil, and the alteration of the soil structure, together mean that a rainstorm may rapidly wash away the charred vegetation and surface soil. Any cracks in the soil can widen and allow further erosion of the landscape.

Nor did I expect to see evidence of the fire beyond the area that was burned. We drove to a picnic spot away from the fire-affected area. On the banks of a stream running through the site was abundant charcoal, as well as some in the channel itself. The charcoal had evidently been washed off the burned area and into the stream, which carried this evidence of the fire tens of miles away. Whole river channels well outside the fire area were filled with a mixture of sediment and charcoal from this sudden flood, and a waterfall was made black with charcoal-laden water.

What became clear during this first trip to a fire-ravaged area was that there was concern not only about the vegetation that had burned but also the impact the fire might have had far beyond the immediate area of destruction. Large quantities of water and sediment can be washed into rivers and cause flooding downstream. Some of this concern was driven by insurance companies which had to pay out for flood damage from well beyond the fire area, in some cases up to 100 miles away downriver. Why was the forest service not doing more to prevent the flooding, even if there was already enough to cope with after the fire?

The Hayman Fire, I would argue, is one that changed our understanding of large fires and their aftermath. So much attention is given to the cause of a wildfire that we tend to forget there are many more factors that play a role in the spread of a fire. In the case of the Hayman Fire, the Front Range of the Rocky Mountains in Colorado had been dry for some months, and indeed generally dry conditions had prevailed for several years. In addition, the fuel loading on the ground was high, partly as a result of previous fire suppression in the forest, which included large areas of ponderosa pine. It was the 'perfect storm'. When the fire started, conditions were very windy. A low pressure system centred over eastern Washington at the time brought strong gusting winds across the area that followed the topography. In the first day the fire spread over an area of 60,000 acres.

We now believe that the fire originated from an abandoned campfire on the afternoon of Saturday 8 June 2002.[5] It began, as usual, as a *surface fire* (Figure 4), burning and consuming the vegetation on the forest floor. The quantity and condition of the fuel meant that it soon migrated into the tops of the trees and became a *crown fire*.[6] Glowing embers were entrained in the smoke plume that was driven by the strong winds and these set off new fires some distance away—a feature called spotting. Despite a quick and aggressive response by firefighters, who employed a number of techniques to control the fire, including the use of air tankers and helicopters, it spread rapidly, so that within a few hours several hundred acres of forest were ablaze.

Overnight, the weather remained dry and warm, and by the following morning the area affected had increased by a further 1,000 acres. The situation was made even worse by the strengthening winds of up to 50 mph and low humidity, allowing the fire to spread through a range of vegetation types. The next day saw a further,

Ground
fuels
} Foliage
Limbs
Logs
Brush
Grass
Duff
Roots

Aerial
fuels
Foliage
Branches
Snags
Moss

Organic
layer
Mineral
Soil

Figure 4. The various types of vegetation fire: (a) *surface fire*, in which only the dead litter and ground cover plants burn; (b) *crown fire*, in which the fire travels via ladder fuels to the tops of trees; and (c) *ground fire*, in which the fire burns the organic layer in the soil, including peat.

large increase in the fire front. Driven by strong winds, it travelled up to 19 miles along the South Platte River corridor towards the Cheeseman Reservoir, a dammed lake in the area that serves as the major water supply for the large city of Denver. The smoke from the fire created its own weather conditions, with the formation of large cumulus clouds, known as pyro-cumulus, that were able to develop up to 21,000 feet over the fire. At this stage the fire was travelling at a speed of more than 2 mph. By this time, though, it was burning on several fronts, so suppression was a major problem.

Conditions improved during the week of 10–17 June, when the wind speeds decreased and the humidity increased, but not enough to halt the fire's spread. When high winds and lower humidity returned on 17 and 18 June, the fire intensity increased, as did the speed of spread. Fortunately, moist monsoon weather then arrived, stopping any further advance, but even so the fire continued to burn until 28 June. By this time more than 138,000 acres had been affected in an area many miles in every direction.

Emphasis is often placed on the success or not of firefighting activities. But it is often a change of weather that is responsible for the full cessation of a wildfire. It is possible for a fire to have been seemingly put out, only for it to continue to smoulder underground. A change in humidity or increase in wind speed can revive it. The Rim Fire that occurred near Yosemite National Park near San Francisco in October 2013 is a case in point.[7] Although much work was put into extinguishing the fire it was not until the weather changed that it was finally put out.

The Hayman Fire was reported by both the national and international media, and although attention waned after the fire had stopped, it was soon realized that the aftermath of the fire warranted further investigation.

One of the immediate impacts was on the river and reservoir. Ash and charcoal had been deposited in the Cheeseman Reservoir even as the fire progressed. This had a number of effects. It clogged filter systems. Water quality was affected, which was of major concern as the reservoir provides the main water supply for the region. The fine ash deposited on the water surface contained mineral ash and a range of soluble elements that can make water undrinkable without treatment. Elements such as phosphorus can stimulate algal growth that can also remove oxygen from the water. It was soon realized that there were wider, more important consequences if there was significant rain following the fire. Some of the sediment resulting from post-fire erosion was washed into the Cheeseman Reservoir, creating an additional problem for water supply as the volume of the reservoir was further reduced, and these flood events also brought contaminants into the water.

This was not the first fire in the area. Indeed, fires of this kind may occur regularly over hundreds and thousands of years. Walking through the burned areas afterwards, it became obvious that there were a number of features of the landscape that could only have been caused by post-fire erosion and deposition (Figure 5).

Two years later, as part of a workshop on post-fire erosion, a group of us visited seven sites that had burned over the previous ten years, including the Hayman Fire site. What was surprising was the quantity of sediment and charcoal still moving across the landscape two years after the fire. A number of treatments had been tried by the US Forestry Service to stop the movement of sediment. In some areas straw bales had been dropped from the air to create a cover so that water is absorbed rather than running off. In other places dead tree trunks had been felled to block sliding sediment.

Another fire site nearby proved even more interesting. The Buffalo Creek Fire took place in Colorado in 1996.[8] The area was the subject of a major investigation by researchers from the United States Geological Survey Hydrology Division, and they were able

Figure 5. Killed forest after the 2002 Hayman Fire near Denver, Colorado, showing successive layers of transported sediment from previous fires.

to show us around. Here, an enormous quantity of sediment had moved overnight a few weeks after the fire, following the first major rainstorm. A large alluvial fan of sediment was formed and the nearby channels were choked with sediment (Figure 6). Some of this was re-eroded and taken further down the river distributary system. This mass movement, which can continue many months after a fire, has implications not only for the immediate area, perhaps changing the courses of streams and rivers and washing away paths or roads, but also for animals and indeed humans both within and outside the immediate fire-affected area. When you see rivers or waterfalls black with charcoal, it is hard to understand why this has not been recorded more often.

The rapidity with which this charcoal-laden water can appear is surprising. One of my research students was in an area called

Figure 6. Alluvial fan produced after a rainstorm following the Buffalo Creek fire, Colorado, 1996.

Sandhills in Colorado in 2010 when there was a rainstorm over a burned area. Within hours, a charcoal-laden stream appeared, apparently from nowhere. This phenomenon has been seen in many parts of the world (Figure 7).

Perhaps the most significant impact of such erosion and deposition events in the recent past was seen following the Yellowstone fires of 1988. These took place across large areas of Yellowstone National Park and at the time shocked the world. With 793,880 acres, or 36 per cent of the park affected, a debate ensued about the effect that a policy of fire suppression may have had on the extent and severity of the fires. Continual suppression of fire, it was argued, had allowed fuel to build up so that when the inevitable fire started it was bound to be larger, more intense, and more difficult to extinguish. Not all agree, however. After the fires, here too extensive sediment movement took place across the landscape and into nearby lakes (Figure 8).[9] Yet despite all this recent

Figure 7. Charcoal-rich flood waters after rainstorm, from a forest fire (the Rodeo–Chediski Fire, Apache–Sitgreaves National Forest, Arizona, USA, 2002).

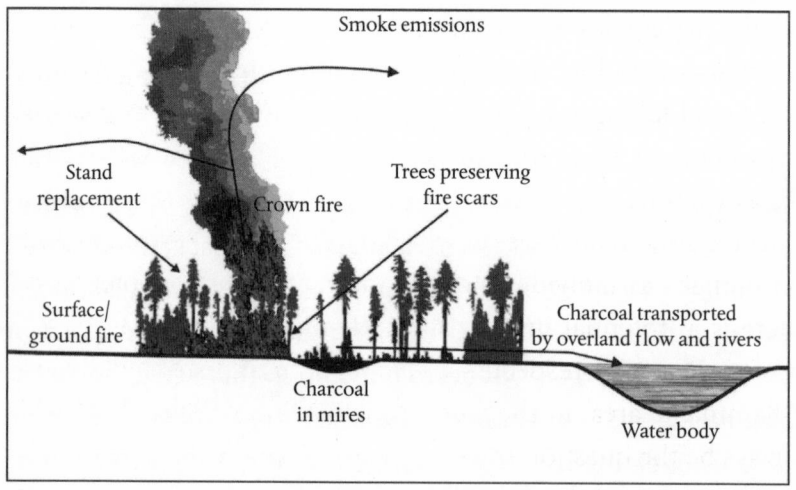

Figure 8. The products of wildfire and their transport.

evidence of post-fire erosion and the subsequent sedimentation, there was still little or no mention of fire as an agent in textbooks. It has taken time to recognize the importance and wider impact of this force of nature.

The toll on communities

The impact of wildfire on human populations is naturally the aspect highlighted by the media. News reports concerning wild-fires focus on how individuals have been affected—through death and injury, or more often the destruction of property. The emphasis is on the drama, and if and when the fire has been extinguished. Rarely, if ever, is there any discussion as to whether a fire should be left to burn, or recognition that it may have been foolish to build houses in a particular flammable landscape. This was amply illustrated by the Black Saturday wildfires of 2009 in Victoria, Southern Australia, which covered 1.1 million acres. Other recent fires have been bigger still. The Fort McMurray fires of 2016 in Canada covered 1.5 million acres, while the 1983 fires in Kalimantan, Indonesia, were bigger still, covering 9 million acres, and killing 173 people. While it was tragic that so many lives were lost, there still appears to be little general understand-ing of wildfire.

There are several stages at which a wildfire may have an impact on humans as individuals or on communities. Most obviously there is a potential impact during the spread of the fire itself (Colour Plate 4). Irrespective of the wisdom or not of building in a flammable area, in the face of a spreading wildfire there will always be the question of whether to stay and defend or to flee. Some fires move slowly and may be restricted to a surface fire. But others burn through the crowns of trees and move quickly, especially if there is significant wind. The rate of spread of a fire can be quite surprising, and fires may be spotted, with new fires started simultaneously in a number of places from glowing embers transported on the wind. This is not only a problem for householders but also for firefighters, as a fire may start unexpectedly some distance away from the main fire front and cut off any retreat. This is what happened in 2013 in a fire near Phoenix in Arizona, the Yarnell Hill Fire, where 19 City of Prescott firefighters died as they were overrun by fire on multiple fronts.[10]

Even if property is not occupied, and even if the most thoughtful defences are put in place, there is no guarantee against wildfire destruction. My friends' house near Boulder, Colorado is a case in point. Their approach had been used to educate the local population about the idea of defensible space: they had thinned the vegetation around the property to prevent any wildfire reaching the house. In 2010, while they were away from home, what came to be known as the Four Mile Canyon Fire took hold in an area not far from Boulder. You can imagine the surprise of the family when they saw on the television news their house on fire while the surrounding trees were unaffected. The main fire did not reach their house and their protective measures had helped with this. But little could stop a glowing ember blown on the wind from setting the place alight.

We have been concentrating on the fire itself and its destructive capability, but we should not forget the role of smoke and other products coming off the fire (Colour Plate 5). Smoke plumes from fires can stretch across hundreds of kilometres, and as we have seen are often visible from space. We can track the spread of a smoke plume both horizontally and vertically. This is important, given the role of smoke as an irritant, and the danger it poses to those suffering from respiratory conditions such as asthma. Smoke plumes from peat fires in Indonesia (Figure 9)

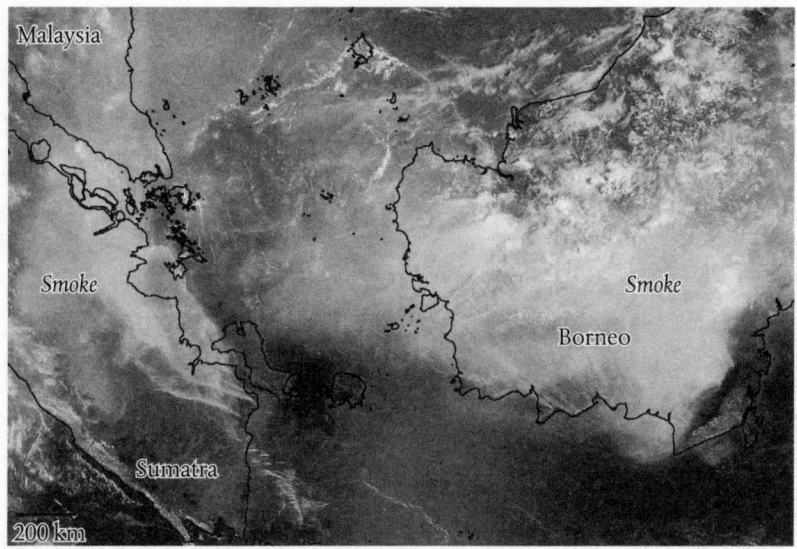

Figure 9. Smoke from Indonesian fires as seen from space, NASA Earth Observatory, 24 September 2015.

have caused extensive smoke pollution in Singapore and Malaysia, and smoke from wildfires in Siberia caused major pollution as far away as Moscow in 2012. Recent research has linked many deaths to the occurrence of smoke from wildfires.[11]

The danger to people is not over with the quenching of the fire itself. As we have seen, intense rain following a fire can result not only in post-fire erosion, but potentially in extensive flooding that can occur many miles away from the wildfire site.

Fire impacts not only people but the buildings and infrastructure of communities, and the question of whether the lives of firefighters should be placed at risk in such cases is a matter of sensitive political debate. Sometimes major infrastructure has been threated by prescribed burns (fires set deliberately to reduce fuel loads in a forest) which got out of control. Such a fire threatened the Jet Propulsion Laboratory in the USA a few years ago (the Station Fire of 2009).[12] Fortunately the fire was controlled before the laboratory was destroyed. Another major incident occurred around the Los Alamos nuclear research facility in New Mexico (the Los Conchas Fire of 2011).[13]

Fires don't just affect humans. We also need to think about the effects on animals and plants at the time of the fire as well as after, sometimes many years later. Animals have two main reactions to wildfire—to flee or to hide. I always expect to see more dead animals following a fire. There are some, and these often appear in emotive photographs. Yet animals can often sense an approaching fire and move away. Often they move towards water, which is safer. One of the most common images is of deer in water with fires raging in the forest behind (Colour Plate 6).

Large animals can usually move quickly enough to escape the fire, but smaller animals may struggle. Many of them burrow or hide in holes, even burying themselves deep in the litter on the forest floor, which may survive the fire. Inevitably, some will not make it. It is not uncommon to see charred beetles and other small insects on the burned surface after a fire has passed. Rare amphibians and reptiles, especially snakes and lizards, are often a cause for concern, not just because they may be killed by the fire, but due to the destruction, even temporarily, of their habitat and food. But even in the largest fires not all vegetation is necessarily burned. Many fires show a mosaic pattern of burning, with patches of unburned vegetation among areas that have been badly affected. These untouched islands can be the source of regeneration and regrowth for communities.

Fire and vegetation

The interaction of fire and vegetation can be very different, depending on the type of vegetation that is affected. There are clearly some types of vegetation that are very sensitive to fire. These often occur in regions such as tropical rainforest, where natural fires are rare. In these areas, many of the fires are started intentionally or unintentionally by humans or human activity. The recovery of the vegetation after a fire in such a region may take many decades or even hundreds of years. In other types of vegetation fires may occur sporadically on a timescale of every few hundred years or so. If left, vegetation may eventually recover but go though a series of changes before the original type is restored. In areas where there is more frequent burning, fire may be less destructive. In such areas fuel loads on the forest floor may be kept down by cool, slow-burning fires in which the main plants are not killed. Even in heather heathland, such as found in parts of the British Isles, the heather may not be killed by a cool surface fire, even though all looks lost in the blackened landscape following the fire, as happened near my house in Surrey, England. Green shoots can appear soon after the first rain, and some areas will revert to the original heather heathland within only a few years.

Some types of vegetation, especially coniferous forests (Colour Plate 7), burn even more frequently, sometimes several times in a century. Such is the ponderosa pine forest of the western US. Here regular surface fires keep down the fuel loads. Without forest management, downed branches and twigs can build up, so that when a fire does eventually take hold it is hotter and more intense, and can climb upwards, creating a much more dangerous crown fire (Colour Plate 8). As we shall see, changes in forest management practice over the past century, public misunderstanding of the nature of fire, and fire suppression, as well as climate change, have all led to an increase in what have been termed 'megafires'.

If a fire does not completely destroy the vegetation, the original plant community structure may be maintained. However, if a fire is severe enough, the tree cover may be destroyed, opening up the area first to invasive or pioneering species and then, if there isn't another fire, a series of more complex plant communities—an ecological succession—until the climax community is reached. If, however, fires are frequent, the climax forest may not reappear.

Adapted for fire

Some types of vegetation are naturally fire-prone, and over millions of years some plants have developed strategies to cope with fire, while others have even evolved to take advantage of fire. A good example of a survival strategy is the development of a thick, fire-resistant bark (Figure 10). This has been achieved by some pine species and, as we shall see later, the trait evolved during the high-fire world of the Cretaceous Period, some 100 million years ago (myr). One of the best-known examples is the thick bark of the giant redwood (*Sequoia*). Fire scars on felled trees show that the trees had survived many fire episodes during their lifetime, which is sometimes more than a thousand years. The thick bark forms a layer of insulation that reduces the transfer of heat to the growing cells in the cambium, the layer beneath the bark on the outside of the wood. But the heat may also cause problems for the water in the xylem cells that transport it from the roots to the leaves.

A number of plants spread clonally, by vegetative growth via underground root systems. They may seem to have been killed by fire above ground, but their root systems can survive and later resprout. Such a mechanism is seen not only in grasses and some shrubs, but also in some trees. The 'quaking aspen' (*Populus tremula*) can often be seen in fire-prone areas in the western United States. It appears as clumps of trees, but these are in fact a single plant connected underground.

Figure 10. Fire scarring but not killing a tree.

Some plants have buds that are protected by bark and which may sprout following fire. This is typically found in some eucalypt trees in Australia, while certain conifers have cones that only open under the heat of a fire, allowing seeds to be shed on to bare soil where there are no other competitors. A good example is the bristlecone pine. There are other plants that release their seeds following a fire, such as some proteaceous plants in both South Africa and Australia. Even more remarkable are the plants that respond to chemicals in the smoke from fire. In the Fynbos vegetation from South Africa there are some plants that release their seeds just as the fire passes, ready to sprout. Hundreds of plants have adopted this strategy. Other plants in the Fynbos have seeds that are stored in the soil awaiting the heat of a passing fire, which triggers their germination.

In some areas, certain plants are commonly found following a fire. The fire weed (*Chamerion augustifolium*), commonly seen in the Rockies, is one example. Another is the blanket flower, or *Gaillardia*, a fire-dependent species whose seeds in the soil germinate only after a fire. The flowers are associated with a moth, *Schinia*, that is often seen on the flower and is dependent on it (Figure 11).[14] Both may be endangered by fire suppression.

Figure 11. The blanket flower (*Gaillardia aristata*), Colorado, USA. Often called 'fire weed', as it blooms after a wildfire has been through an area, the flower is a fire-dependent species. The seeds only germinate in the soil after a fire has passed by. Associated with it is the well-camouflaged blanket flower moth (*Schinia masoni*). Both flower and moth have become rare because of fire suppression.

Certain vegetation types are well adapted to frequent and, in some cases, hot fires. Shrubs of the Chaparral in southern California burn in high-intensity crown fires. Another example can be found in central and northern Australia. Here a tussock-like grass called spinifex (*Triodia*), burns every few decades. Other plants in this type of vegetation also have adaptations to fire.

Savanna grasslands in Central and South Africa are made up of drought-resistant grasses known as C4 grasses. This savanna needs to burn. The grasses have found ways to resprout readily following a fire. Plants in such fire-prone areas may follow a strategy proposed in evolutionary ecological theory called 'kill thy neighbour'.[15] If a plant burns but is not killed (it may have a number of fire-adapted traits) while its neighbours are killed as the fire spreads, this leaves the survivor space to spread into the now vacant surroundings. All these fire-associated adaptations and strategies point to a long evolutionary past of interaction with fire.

Controlled fire

In many parts of the world the use of fire is prevalent, and it is not a force to be feared. But over the past hundred years or more there has been a shift in attitude to fire. Stephen Pyne, an expert on the history of fire, has argued that the move into urban centres has had a profound effect on our relationship with fire. He cites our need to both harness fire for energy, heating, and transport, and to exclude it from buildings, describing this shift as the 'Pyric transition'.[16] We shall come back to this idea at the end of the book.

As more people move into urban centres, where fire is something to be excluded, the prevailing attitude of the majority tends to be that all fire is bad. The reaction to a wildfire may be extreme, and the response ill-informed. An understanding of the role of fire and its relationship to life over the long history of the planet provides the necessary wider perspective, and even more so as we grapple with the impact of climate change.

To begin our exploration of the history of fire, we need to look first at one of the main clues to fires past—charcoal.

2

Getting dirty:
what charcoal can tell us

Most of us are familiar with charcoal from sketching with it at school, or using charcoal bricks for a barbecue. You will have noticed that it got your hands dirty, that it is brittle, and that it is quite light—at least, lighter than an equivalent piece of uncharred wood. You may also have associated the black residues left after a bonfire with charcoal. If you have been to an area where the vegetation has been destroyed by wildfire, you may have also noticed black residues of charcoal on the ground that make a crunching sound beneath your feet (Figure 12).[1] Our first two examples of charcoal are both products of human manufacture. The bonfire charcoal is a naturally formed material, but still the link with wildfire may not be made. When we see images of burning vegetation it is natural to imagine that all the plant material is consumed by the flames. Yet, as I came to realize on my visit to the site of the Hayman Fire, there is often a significant quantity of unburned material, and charcoal residues as well.

Making charcoal

Why are we left with charcoal after a fire? Charcoal is produced by heating plant material (most commonly wood, but not exclusively

Figure 12. Charcoal produced after surface fire, Frensham, Surrey, 1995.

so) in the absence of oxygen. So it isn't a product of the fire itself, but of the intense heat from the fire.

Wood is essentially made up of two organic compounds: *cellulose* and *lignin*. Both compounds consist of carbon, hydrogen, and oxygen, but they differ in structure and therefore in properties. In cellulose, the carbon atoms are arranged in straight lines (it is an example of an *aliphatic* compound). It is the material from which paper is made. In lignin, on the other hand, the carbons are arranged in rings (it is an *aromatic* compound), and it is this structure that gives wood its toughness and strength.

Industrial charcoal is used for a variety of metallurgical processes, and as adsorbents and food additives, as well as for barbecues and artists' materials, so its formation has been carefully studied. As temperature is increased, a series of physical and chemical changes occurs in wood. These involve the breakdown of cellulose and lignin into a variety of compounds. Initially, the wood absorbs heat, and from 20 to 110°C, water is driven off—it becomes bone dry. From 100 to 270°C the remaining water, some of it bound to the chemicals that make up the wood, is driven off. At this stage the wood also starts to decompose, giving off gases such as carbon monoxide (CO), carbon dioxide (CO_2), and methane (CH_4), as well as acetic acid (CH_3COOH), methanol (CH_3OH), and other compounds. At the next stage, from 270 to 290°C, heat is still being absorbed, and some tar, a black-brown mixture of liquid hydrocarbons and free carbon, may be produced; the release of these complex compounds increases as the temperature rises further. Above 290°C some exothermic reactions (reactions that give off heat) start to take place, further increasing the temperature. This is the threshold at which the formation of charcoal begins, through a process called *pyrolysis* (technically, heating in the absence of oxygen, causing a thermochemical decomposition of organic materials at high temperatures). By about 400°C to 450°C, the process of charcoal formation is practically complete. But there is still an appreciable amount of tar trapped in the organic structure of the charcoal, so above this temperature more of the tar is driven off until about 500°C, when good commercial-grade charcoal is produced. The charcoal remains a solid material to well above heating temperatures of 1,000°C or more, albeit more fragile and brittle.

How then does our charcoal compare with the material from which it was derived? Physically, the most noticeable feature is a change in colour from brown to black. The charcoalified wood, or charcoal as it is now, is also much more brittle. It can be broken between the fingers with a small amount of pressure. This easy crumbling is also why a black streak is left when a piece of charcoal is drawn across paper, and why it marks your fingers. And it is lighter than the original wood, having lost mass during the charcoalification process. However, as we shall see, the anatomical integrity of the cells remains.

Another feature is noticeable on partially charred wood: it will often show not only blackening but cracking. These cracks can form a grid pattern on the surface of the wood and, as shrinkage continues, centimetre-sized cubes of charcoal remain. These can be seen in any residue after a fire (Figure 13). They are typical of material formed in wildfires and may also characterize fossil charcoal.

Chemically, during the heating process, the hydrogen and oxygen of the cellulose and lignin making up the wood are driven off and the cells become enriched in carbon. As the temperature

(a)

(b)

Figure 13. (a) Partially charred log; (b) detail of charcoal showing small stems and cubes of wood from the fire in Figure 12.

increases these carbon atoms become more ordered. The result is that the material can act not only as an absorbent, but also becomes resistant to decay. There is no reason to suppose, therefore, that charcoal should not be abundant in the fossil record. And it would be evidence for fire in the geological past.

The recognition of fossil charcoal

I had thought little about charcoal during my undergraduate years studying geology. I was far too excited looking for 'real fossils'—marine shells, sharks' teeth, and the like—to take any notice of black fragments in the rocks. What surprised me most when I did start to become interested in fossil charcoal, more than 40 years ago, is how few records of charcoal there were in the scientific literature.

In the seventeenth century, Robert Hooke made many telling observations regarding both charcoal and fossil wood. He argued that fossil wood arose from wood that had first been buried and then became petrified, challenging the contemporary view that it was formed from stone. Hooke conducted experiments involving charring plants and recorded his observations in his great work, *Micrographia*. He was able to show that charcoal was formed through the action of heat, while the presence of air was needed for combustion, or flame, to occur.[2] But Hooke did not encounter charcoal in fossil form.

One of the earliest records of fossil charcoal came from the writings of the great Victorian geologist Charles Lyell (Figure 14). In a paper in 1847 he reported charcoal occurring in the coalfields of South Wales as well as eastern Virginia (Figure 15).[3] At this time it was often called 'mineral charcoal', and its significance was not fully appreciated.

Figure 14. The young Charles Lyell.

If fossil charcoal was recognized as early as the middle of the nineteenth century, why was it recorded so rarely, and why was its significance not realized? A clue comes from the use of the name 'mineral charcoal'. This was translated into French and German, where it became *fusit* or *fusain*. It was Marie Stopes who introduced fusain into English.[4] The widespread use of the term fusain, therefore, took away its identification with charcoal, and its origin came to be increasingly disputed.

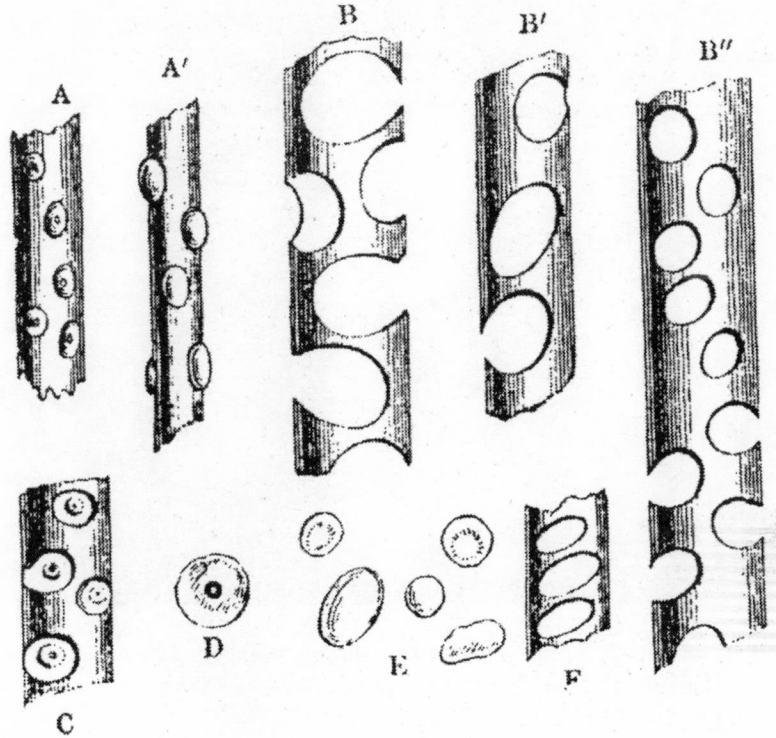

Figure 15. Illustration of Carboniferous (320 myr) charcoal by Charles Lyell, 1847. The wood cells (tracheids, which transport water in the xylem) show that the pitting in the cell walls has been preserved.

Marie Stopes (Figure 16) is probably best known today for her work on birth control and her book, *Married Love*.[5] But she was a significant scientist. Stopes had written an important paper on the origin of 'coal balls'—unusual round stones found in coal seams—so it was not surprising that, as coal was the major source of power during the First World War, she was asked to become involved in studies of British coals in a laboratory set up in 1916 under the aegis of the Department of Scientific and Industrial Research and headed by R.V. Wheeler. This work resulted in an important monograph, published by Stopes and Wheeler in 1918, on the *Constitution of Coal*. Stopes went on to

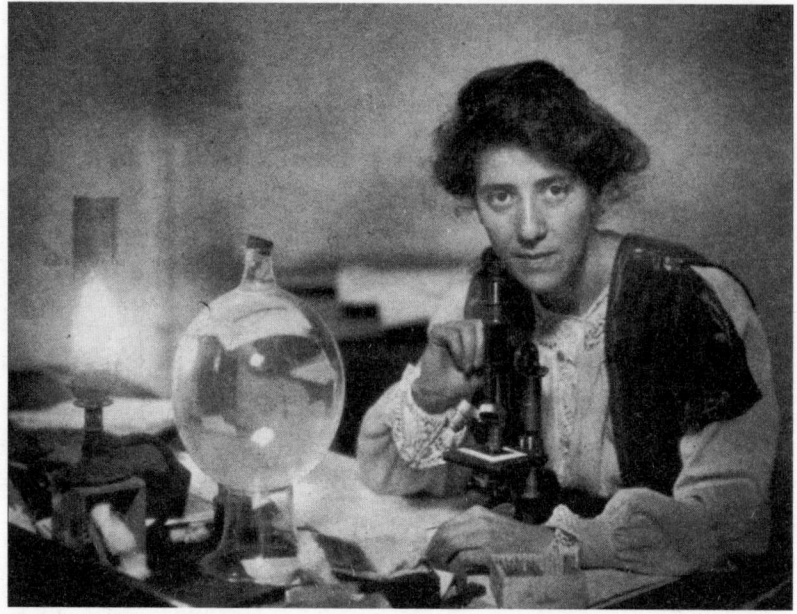

Figure 16. Marie Stopes, the family planning pioneer, was also a scientist who made major contributions to the study of fossil plants and coal geology.

publish her own observations, and establish a system of nomenclature for coals.

Marie Stopes described fusain as reflecting light well, having a silky lustre, and having a chemical composition of almost pure carbon.[6] By the beginning of the twentieth century there was increasing uncertainty about what fusain represented, though some scientists did believe that it was charcoal derived from forest fires.

One of the main problems was that fusain was found to occur in coals. These formed in the Carboniferous Period, 360 to 300 million years ago (see International Geological Time Scale in Appendix), when peats that developed under ever-wet conditions, with significant rainfall every month of the year, were metamorphosed by burial and heating in the Earth's crust. The widespread occurrence of fusain in coal was already known by the 1880s. But charcoal is much less common in modern peats than in some ancient coals, and many thought fire could not occur in wetland settings. Coals are fossil peats, and we know that peats can only form in ever-wet conditions. Occasionally the peat surface may dry, and fires may spread across the peat. This had been well demonstrated in the peats of the south-eastern United States.

Others argued that the preservation of delicate fern leaves as fusain made the idea of fire untenable. Not all agreed. A lone voice in the 1950s was that of Tom Harris, who in 1958 published a paper on forest fire in the Mesozoic, describing fossil charcoal from deposits in south-west England and southern Wales of early Jurassic age, around 200 million years ago.[7] This was based on the discovery of wood and leaves of conifers preserved as fusain in these deposits.

The debate rumbled on well into the 1960s. My doctoral supervisor Bill Chaloner, who knew Marie Stopes, told me that she would never accept that fusain represented charcoal formed by wildfire. The use of the term fusain and the increasing numbers who began to doubt that it represented charcoal made even recording the material superfluous.

Despite the significance of fusain highlighted by Harris's 1958 paper, it was largely ignored. His former research student, Ken Alvin, discovered beautifully preserved charcoalified ferns from the Cretaceous Wealden rocks of the Isle of Wight, some 130 million years old. He had no problem believing they were the result of ancient fires, and Harris subsequently described fossil charcoalified ferns of the same age from southern England.[8] You would imagine that with this growing evidence fusain and charcoal would be considered the same material, i.e. charcoal derived from wildfire. Yet this was not the case when I started my own research in the early 1970s. Some influential scientists, such as the American palaeobotanist Jim Schopf, continued to maintain that fusain could not be charcoal from ancient wildfires, and that such excellent preservation was unlikely from a 'sudden thermal cause'. Some researchers wondered whether fusain might be produced by another means, such as oxidation of the surface of the peat.

It became clear, when I started to find fusain commonly not only in coals (Colour Plate 9) but also in sediments, that I needed to address the issue of the equivalence of fusain and charcoal. To convince the doubters, I would have to compare the two materials in a wide variety of ways. I examined the physical and chemical characteristics of fusain, and, unaware that Hooke had done the same as far back as the 1660s, I conducted some charcoal experiments of my own. I made charcoal in my own garden bonfire, and used my mother's oven to char conifer leaves. I was then able to compare these under the microscope with fossil fusain that I had collected.

Charcoal under the microscope

Robert Hooke was the first to make observations of the cellular structure of wood charcoal (Figure 17).[9] Later, Charles Lyell, in his observations of mineral charcoal from the Carboniferous, showed that the wood fragments had preserved anatomical details (Figure 15).[10] But it was not until the latter part of the twentieth century that a breakthrough was made with microscopic observations on fossil charcoal.

The main reason for this breakthrough was the invention of the *scanning electron microscope*, or SEM, that allowed high magnifications

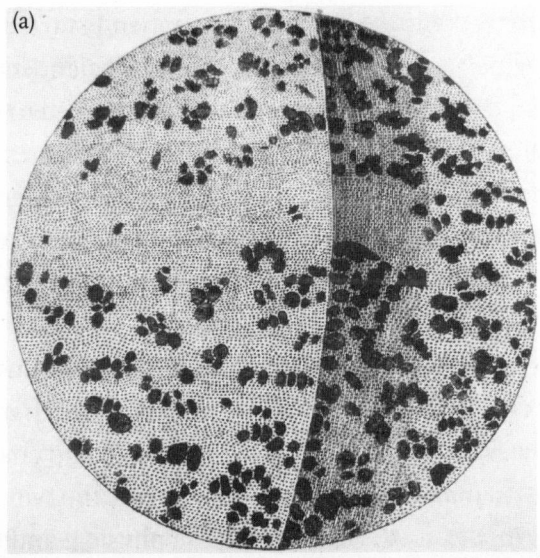

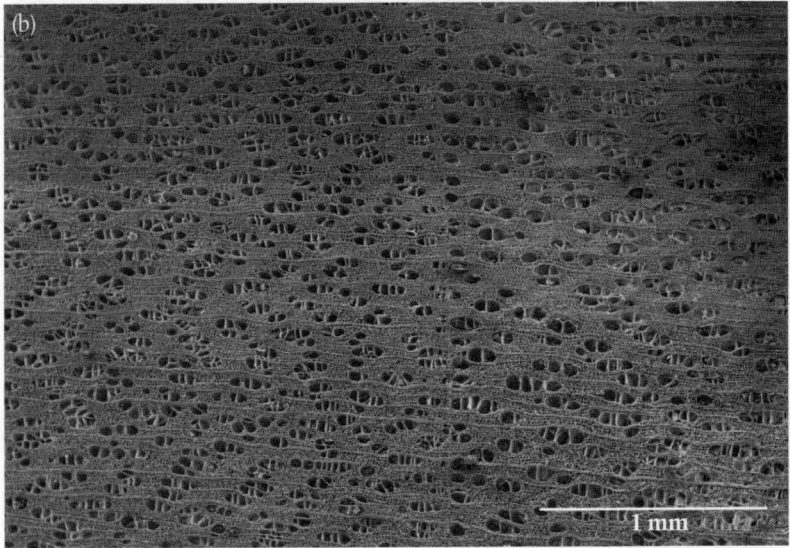

Figure 17. Charcoal illustrated by Robert Hooke in his work *Micrographia*. (a) Hooke's drawing of the charcoal; (b) a modern scanning electron micrograph of a similar piece of wood illustrated to the same scale.

of objects to be obtained by knocking electrons off the surface of a sample, which are collected and analysed to produce a three-dimensional image of the surface. Until the 1960s, most high-magnification microscopy involved mounting a very thin polished sliver of rock on a glass slide and shining light through it. Such 'thin sections' don't work well for charcoal because the cell walls are black and brittle. The SEM made the need to section a specimen unnecessary, and the technology easily provides magnifications from 40 to many thousands of times, which is sufficient to see the relevant details of charcoal. The resultant three-dimensional images are often striking (BW Plate 1). In one of the early papers showcasing this new form of imaging, published in 1976, the authors demonstrated the change in structure to plant cell walls that occurs during the charcoalification process.[11] Wood cell walls have a layered structure with a central thin layer (the 'middle lamella') of a polymer, pectin, between adjacent cell walls. During charcoalification, this layering is lost, and cell walls become homogenized. Soon, excellent images were also being obtained by SEM of ancient charcoal.[12]

SEM imaging was ideal for my investigation of fusain, allowing detailed comparison of its three-dimensional structure with that of modern charcoal. It was now possible to demonstrate that fossil wood charcoal showed homogenized cell walls, just as seen in modern charcoal. Yet there were still many who did not accept that fusain represented fossil charcoal, in particular because of the occurrence of fusain in some coals.

Just as rocks are made up of minerals, Marie Stopes conceived that coal was composed of *macerals*. She devised a classification of coal macerals, now known as the Stopes–Heerlen system. Two of the macerals of the 'inertinite group', called fusinite and semi-fusinite, are both found in fusain occurring in coal. Although making thin sections of coal for microscopy by shining light through was difficult, a method was devised in which the coal was embedded in resin and polished. The polished block would then be observed under oil, using a reflected light microscope. This enabled the cellular structure of the fusinite and semifusinite to be observed. The cells often showed shattered cell walls with angular breaks—a feature called *bogen-structure* (Figure 18). The resin technique also allowed the measurement of quantitative data from the polished surface, in particular its *reflectance*, the proportion of light that the surface reflects. The inertinite group of macerals are all characterized by high reflectance when observed in this way. The difference between fusinite and semifusinite is

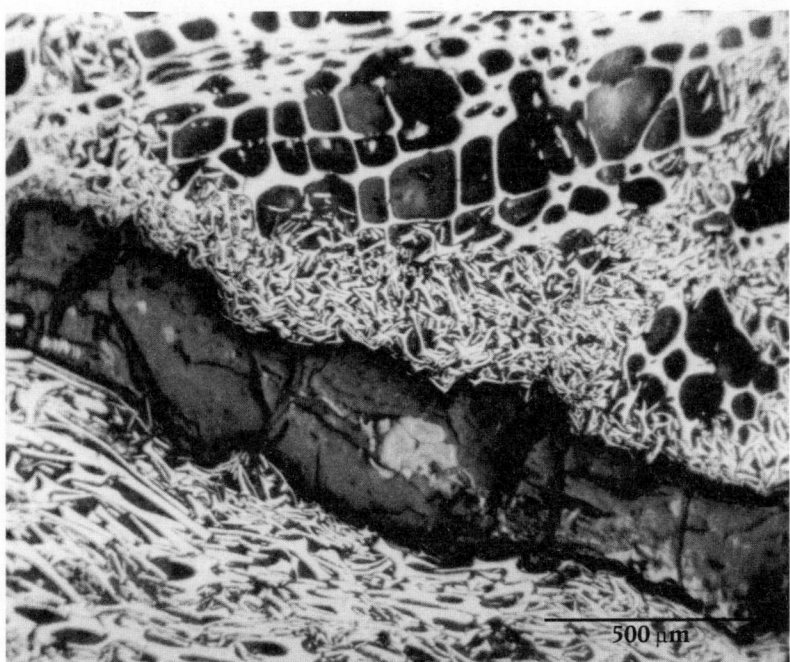

Figure 18. Fusinite (fossil charcoal) in a Permian (290 myr) coal from Australia showing whole cells and crushed cell walls known as 'bogen-structure'. The dark material is uncharred coalified wood known as vitrinite.

mainly in their reflectance value: fusinite is much brighter—that is, it has a higher reflectance.

The fusain found in sediments proved to have similar characteristics to fusinite and semifusinite in coals. I was able to show that the fossil fusain found in sediments had high reflectance and would be called fusinite and semifusinite if found in a coal. Charcoal also has high reflectance. This characteristic appears to be a result of the charring process (and not of the surface oxidation of peat, as had previously been claimed for fusain). And another interesting fact emerged from experiments with charcoal: the reflectance of the charcoal was found to increase with charring temperature (Figure 19). And the increase in reflectance

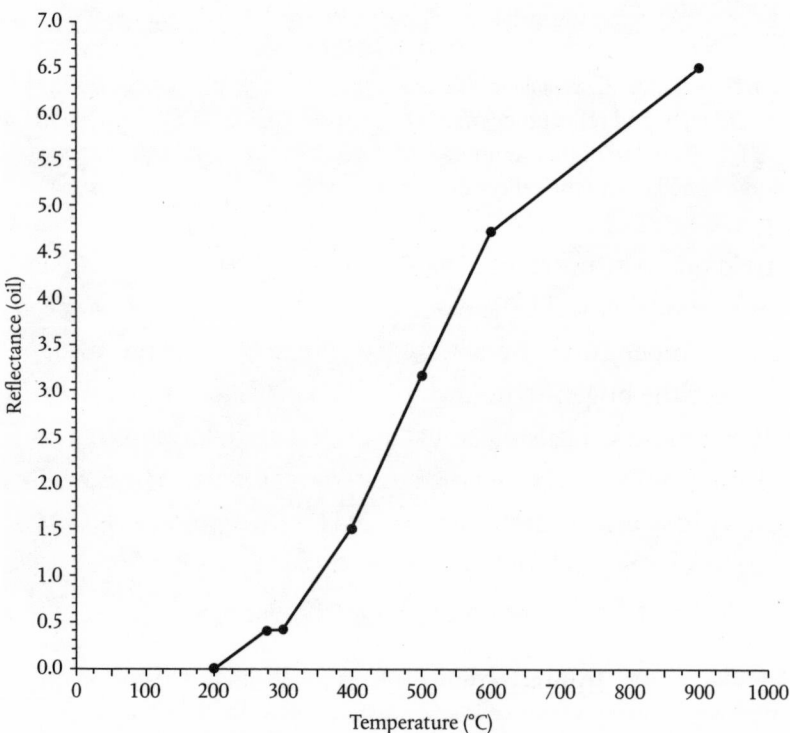

Figure 19. Increasing reflectance with temperature of *Sequoia* wood experimentally charred for one hour.

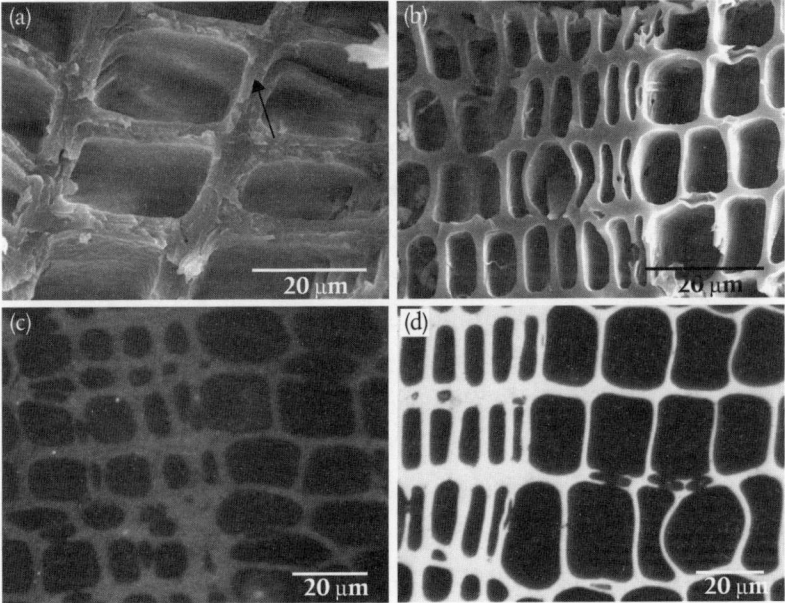

Figure 20. Charred *Sequoia* wood, 350°C (a, c) and 450°C (b, d), seen under the SEM (a, b) and reflectance microscope (c, d). The cell walls homogenize when subjected to higher temperatures (the arrow in (a) indicates the fused middle lamella), and the reflectance increases.

turned out to be correlated with the fusing of the middle lamella of the wood cell walls (Figure 20), and their breakdown at even higher temperatures. By comparison, then, it becomes evident that both the bogen-structure and the high reflectance observed in fusinite and semifusinite must have occurred during a charring process, prior to burial and metamorphosis into coal. By now, the identity between modern charcoal and fossil fusain was clear, and there was no reason not to call the fossil material charcoal.

Investigating fossil charcoal

Establishing the identity of fusain and charcoal was important, because researchers had begun to suspect that as charcoal was so inert it would easily be found in the fossil record. Further, the occurrence of charcoal in the fossil record would be able to tell us about the occurrence of fire through time, and the role fire may have played in plant evolution. Palaeobotanists—those who study fossil plants—came to realize that not only was the anatomy of individual plants well preserved in fossil charcoal, but that these data could tell us something of the vegetation that was being burned, and of the landscape and climate of the time.

Any plant part can be preserved as charcoal. My own first discovery was of charcoalified leaves (BW Plate 2), of a plant from the Carboniferous Period, a conifer later named *Swillingtonia*, after the Swillington Quarry near Leeds in Yorkshire where it was found.[13] These conifer leaves preserved the *stomata*—the pores in leaves through which plants exchange gases with the atmosphere. The stomata of these charcoalified *Swillingtonia* leaves were later used to determine the CO_2 content of the atmosphere 300 million years ago.[14] The density of stomata is directly related to the concentration of CO_2 in the atmosphere: simply, plants growing in low concentrations of CO_2 have more stomata than those in high-CO_2 concentrations. The *Swillingtonia* leaves had large numbers of stomata, suggesting a low-CO_2 atmosphere and a cool, 'icehouse' world. So we can glean an idea not only of the vegetation, but also the atmospheric composition and climate of the time, from plants fossilized in charcoal.

Charred leaf material had been described by various researchers since the 1950s, but what brought home the quality of plant detail preserved in fossil charcoal was the work of the Danish palaeobotanist Else-Marie Friis and her colleagues, who described a charcoalified fossil flower from rocks more than 70 million years old.[15] Else-Marie brought material with her to London, and Bill Chaloner and I were quite astounded at the level of preservation of such flowers. Most of us would not imagine that something as delicate as a flower could be preserved as charcoal resulting from a wildfire. Later, I was able to show that residues from modern fires on heather heathlands preserve large quantities of charcoalified flowers.[16]

If we were to understand where charcoal might occur in the fossil record, it became important to learn how charcoal behaved during transport. Much was known about the behaviour of charcoal in wind; yet, as I embarked on a postdoctoral fellowship at Trinity College Dublin in the mid 1970s, strangely little information was available on the behaviour of charcoal in water. I was to have a chance to engage with the problem very soon. We took students on a fieldtrip to early Carboniferous rocks on the south Donegal coast, stopping at Shalwy Bay to see the evidence that the sea level rose over 340 million years ago to transform the land into a warm tropical sea with corals. But what struck me was that the first marine rocks in the section were very black (Figure 21). On closer inspection it turned out that these rocks, which yielded typical marine fossils, contained a large quantity of charcoal. I immediately collected some samples to take back to the laboratory, where I dissolved the charcoal out of the rock with acid and examined it under the SEM. What wonderful preservation! The details even of the cell wall thickenings could be seen (BW Plate 3). But as the fire would have occurred on land, how did the charcoal get transported to the sea and deposited in near-shore sands?

A group of my students and colleagues were sent to Donegal to investigate the problem. What they discovered reinforced our need for new knowledge. They mapped out the charcoal-bearing rock layer and showed that it was very widespread, and was a single unit representing tidal sandbars within an estuary that had been inundated by a large quantity of sediment and charcoal brought down by rivers draining off the highlands to the north. This area must have been clothed in vegetation that had burned

Figure 21. The early Carboniferous rocks (325 myr) at Shalwy, Donegal, Ireland, showing black, charcoal-rich deposits overlying tidal sandstones deposited in the sea.

in a wildfire, and the process of post-fire erosion had moved the sediment and charcoal, just as in modern fires such as the Hayman Fire. The material had ended up in a river system that drained into the warm Carboniferous sea. The charcoal appears to have been the result of a single wildfire event, and calculations showed that the fire must have been very large—larger than the combined area of Holland and Luxemburg today (Figure 22).[17]

How far had the charcoal travelled? Answering this question proved more difficult than it would seem at a first glance. We needed to know how long charcoal takes to settle in water. Wood charcoal changes in size during the charring process, and the alterations in cell wall structure at different temperatures may affect its floating/sinking behaviour. The only way to investigate was to do experiments with charred wood in a wave tank. We found that although charcoal formed at low temperatures sank

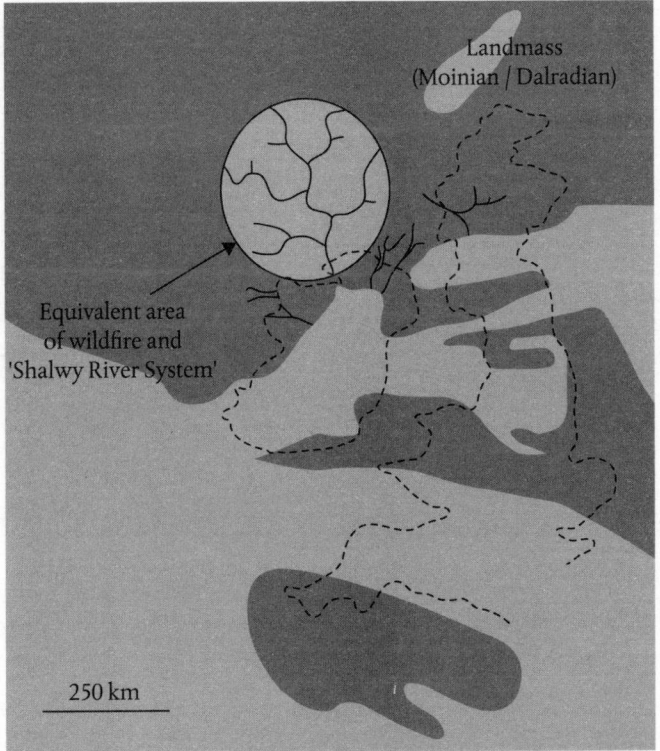

Figure 22. Reconstruction of the British Isles 325 million years ago, showing the extent of the fire and drainage area. Dark areas are land.

much like dead wood, that produced at temperatures around 325°C, when the cell walls became homogenized, stayed afloat for longer. At 600°C the cell wall began to break down, and the resulting charcoal sank more quickly. So charcoal could remain in suspension in water for much longer than had been thought, and could therefore be transported much further than had been imagined. Large pieces of charcoal had been assumed by many to represent a local fire, but now it appeared they could have come from an event a long distance away.

At about this time another event occurred that was to give us new insights into fire and its products. In May 1995, a fire broke out in the Frensham Common Nature Reserve in Surrey, England, close to my house. Indeed, I could see the smoke from my window. I telephoned my colleagues and we arrived at the site just as the fire was brought under control. Walking across the charred landscape even melted the bottom of our Wellington boots! Arriving at the site so soon after the extinction of the fire allowed us to do several things. We were able to collect charcoal before it was dispersed by wind or water. We were then able to observe how the charcoal was moved by wind and water over the next three years. And we could identify the plants and plant parts that had been charred and compare them with the vegetation that had existed before the fire.

The first surprise was that the charcoal left behind was not simply wood charcoal but included all the organs of plants, even small heather flowers. The second surprise was that after a few days the wind picked up the finer charcoal and produced wind ripples of charcoal on the bare land surface, and the charcoal in these wind ripples turned out to contain large numbers of char-coalified flowers (Figure 23). Perhaps this mechanism could account for the occurrence of some of the concentrations of charcoalified flowers found in the fossil record.

Subsequent rainstorms began to move the charcoal into hollows and then streams, concentrating the wood charcoal. This suggested we might explore how different sizes of plants and different plant organs might behave during water transport. So we bought an industrial-sized kiln and started to make charcoal from different plants and plant organs at different temperatures and test them in running water in a flume tank, in which you can vary the rate of flow. We could observe how the charcoal became incorporated into the sediment at the bottom of the tank, and we found that different plant organs, sizes of charcoal, and charcoals formed at different temperatures, all behaved differently from each other.

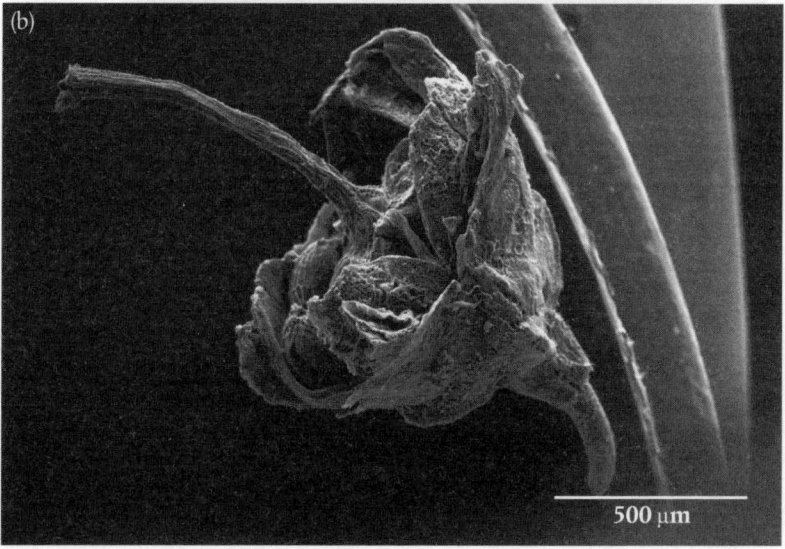

500 µm

Figure 23. (a) Charcoal produced after surface fire, Frensham, Surrey, 1995 and blown into ripples by the wind. The ripples concentrate charred heather flowers that can be imaged (b) using the scanning electron microscope.

So not only could charcoal travel a long way in water, but during transport, different types of charcoal became separated. That would explain why some fossil deposits comprise only wood charcoal fragments of a particular size, and why only some contain small charcoalified plant organs such as flowers or leaves.

The temperature of fires

We had set out to use measurements of reflectance to show the identity of charcoal and fusain—to confirm that it was indeed fossil charcoal. But as research progressed, it became increasingly obvious that the reflectance technique had a number of applications. In particular, the reflectance of charcoal could provide information on the temperatures reached in a wildfire. For example, we were able to show from the reflectance measurements on the resultant charcoal that the Frensham fire of 1995 only produced temperatures of around 400°C to 450°C.[18]

We had, until now, used only wood in our experiments, but would the reflectance of other plant material also change during the charring process? Ferns don't contain wood, and they are often found as charcoal. Our experiments showed that the change in reflectance of ferns during the charring process was the same as for conifers and wood from flowering plants. And we also noticed that the anatomical details of the ferns were retained during charring, and could easily be recognized in the charcoal. What of non-plant material such as fungi? There had been claims that fungi had naturally high reflectance but this was not well established. Bracket fungus is commonly associated with trees, especially dead and dying trees, so looking at such material was potentially useful. Chemically the fungus is made of plant chitin, and not cellulose or lignin as in the case of wood. We found that this and other fungal material had no inherent reflectance and the reflectance of the resulting charcoal rose with increasing temperature just as with plant material. So, irrespective of botanical origin, plant and fungal material show increasing reflectance during charring with temperature, and all could be used to help interpret the temperatures of wildfires.

We already knew that at a given temperature the reflectance rose with time (Figure 19). Most experiments had used times at the highest temperatures for no more than an hour, as work on wildfires suggested that plant material would not experience high temperatures for more than an hour, and often considerably less.

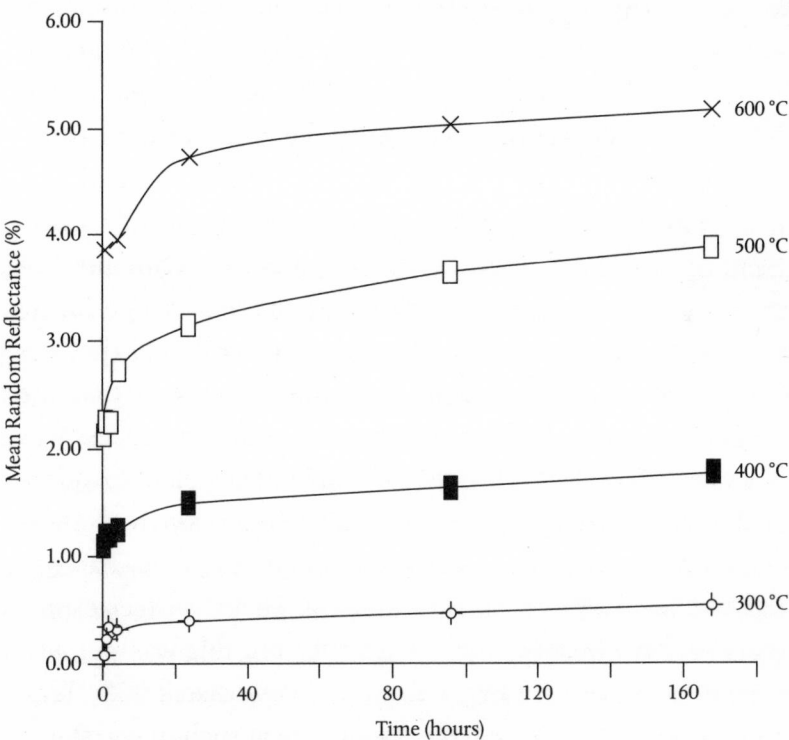

Figure 24. Increasing reflectance with temperature of experimentally charred *Sequoia* wood with varying times.

Reflectance data on our experimentally produced charcoals showed that reflectance continued to rise over four hours of heating, but began to level off significantly after that (Figure 24). So the reflectance data could provide a minimum temperature of formation, even in a wildfire where the time taken for the charring was not known. If, for example, the reflectance value was high, the charcoal could not have been produced at low temperatures, say less than 300°C, however long the plant material was exposed to that temperature. Incidentally, the same technique, applied to charcoal embedded in volcanic pyroclastic flow deposits, which remain at high temperatures for a long time, provides a way of calculating the temperatures reached in such flows—this has proved useful for volcanologists.

New tools to understand the past

For a long time, scientists had to destroy specimens—slicing them up into thin slivers—in order to examine them under the microscope. Today we have techniques involving X-rays, such as CT scans and X-ray tomography, with which we can explore the innards of a specimen without cutting. The X-rays may be obtained from a particle accelerator. The Swiss Light Source is one of the world's leading synchrotrons. Electrons are accelerated in a circular tube, and the bending of the beam results in the emission of pure, single-wavelength X-rays. These X-rays can be focused and used to produce a series of images across a rotating specimen. The images can then be stacked together using sophisticated computer software to construct a three-dimensional image of the specimen, which can be viewed from any angle (BW Plate 4). All this can be done without damaging the specimen—a great bonus when dealing with delicate and rare fossil material.

We used this technique to study some of my oldest charcoal material from the early Carboniferous of southern Scotland. These were charcoalified pollen organs and an ovule that was only 1.5 millimetres long. The SEM could be used to examine the surface of the specimen, and revealed that the embryo had beautifully preserved glandular hairs.[19] But we wanted to investigate any internal anatomy that might have been preserved. The SRXTM (Synchrotron Radiation X-ray Tomographic Microscopical) technique allowed us to see the internal structure of the ovule without destroying it. We were able to reconstruct the ovule, pick out the inner and outer surfaces and the hairs in different colours, rotate the specimen, and strip away the different layers in the image (BW Plate 5). Modern imaging techniques like this have been a great boon for palaeontologists.

We have come a long way from the time of Robert Hooke in 1665. Yet the beautiful preservation of fossil charcoal is yet to be widely appreciated.

3

Kindling

What does it take to make a fire? The factors underlying fire can be illustrated with a triangle, and five fire triangles, relevant to different scales in area and time, have been defined (Figure 25). Let's start with the most basic, at the smallest scale. The 'fire fundamentals triangle' has three elements: fuel, as there needs to be something to burn; heat, because fires can't start without a source of heat; and oxygen, essential for a fire to combust and spread. The importance of oxygen becomes obvious when we put out a fire. The use of sand or CO_2, or even smothering, is a way to exclude air, and more specifically to remove oxygen from the system so that the combustion reaction stops. Water has two effects. It reduces the amount of oxygen getting to the fire, but more importantly the heat energy from the fire goes into evaporating the water rather than heating the fuel that allows the combustion reaction to continue.

Our second triangle can be called the 'fire environment triangle'. Here again, fuel forms one of the points. Another is the weather, as this controls the moisture in the fuel, affecting its flammability. The drier the fuel, the more easily it can burn. Perhaps surprisingly, the third point of the triangle is topography, which impacts on the rate and pattern of spread of the fire. Hill slopes, for instance, can provide an updraft of air that allows the fire to spread more quickly.

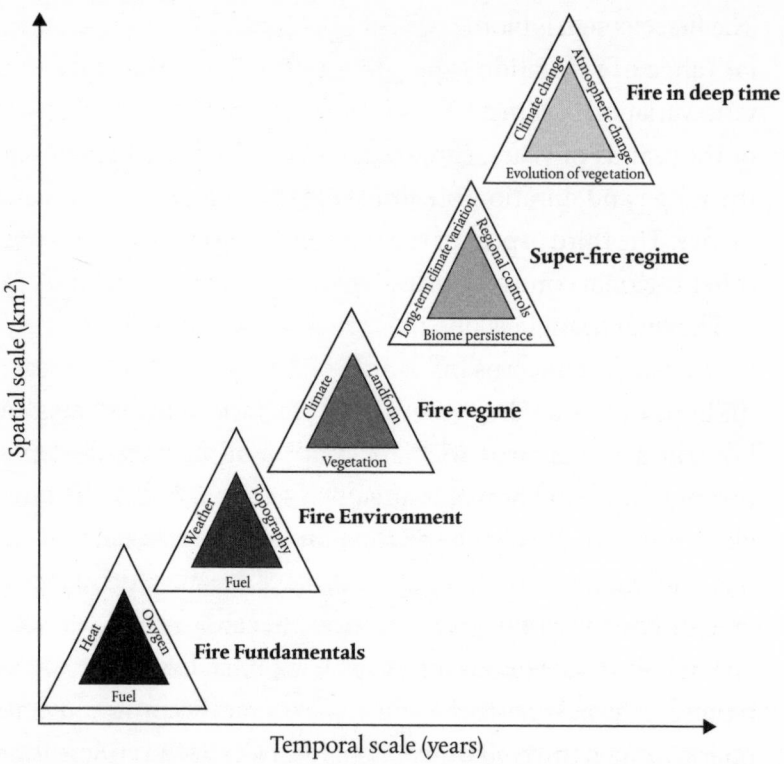

Figure 25. Fire triangles: from the local to the global and through time.

The next triangle up widens our perspective in terms not only of spatial scale but also time. This triangle can be called the 'fire regime triangle'. Here we consider not simply the fuel but the type of vegetation that is being burned. Some types of vegetation are more flammable than others. The overall climate is also significant at this bigger scale. For example, temperate seasonal climates are more fire-prone than wet tropical climates, where there is rain every day. The third arm of this triangle is landform: mountainous regions are more susceptible to fire than low-lying flat areas.

The fourth triangle of fire has only recently been proposed. This is the 'super-fire regime triangle'. Here too, time plays a role.

The first element is biome persistence—that is, how long a particular range of vegetation types might persist. The second is long-term variation in climate. We know, for example, that on timescales of thousands of years, climate can vary. How this happens, and the range and duration of climate change, can all have an effect on fire. The third aspect of the triangle brings together a range of other regional controls, including changes in topography.

The final triangle is one that I first proposed. The 'fire in deep time triangle' concerns influences on the geological timescale of millions of years. The evolution of vegetation forms one side— how the size, construction, and strategy of plants have changed through time, and how this affects the nature of fire. The second element is climate change. On a geological timescale, the world has undergone a number of profound changes in climate, from greenhouse to icehouse worlds, and each climate state has its own effects. The final element takes us back to the first triangle, as it represents atmospheric change, in particular the change in atmospheric oxygen through time. This aspect has become increasingly important to our understanding over the past 10–20 years.

If we are to understand the history of fire in deep time we need to consider the fundamental elements of fuel, heat, and oxygen on geological timescales—vegetation, the source of ignition, and oxygen levels in the atmosphere.

The evolution of fuel

There can be no fire without a fuel to burn, so there would not have been any wildfire before plants evolved on land. There may have been some mosses or algae on the land surface more than 450 million years ago, but it was not until the Silurian Period, around 420 million years ago, that we see evidence of the first recognizable *vascular land plants*, with the ability to transport water and nutrients through *xylem*. These were to give rise to most of the plants we are familiar with today.

The first plants were small herbaceous forms that produced spores. They were only a few centimetres tall and lived in small patches near water, but were not sufficiently abundant to form a significant fuel to allow a sustained wildfire. Through the next 50 million years, into the Devonian Period, plants continued to diversify, but although some became larger, and so could provide more potential fuel to be burned, they were still small, reaching only one metre in height at most, and reproduced by spores that limited their distribution to wetter environments.

Two significant developments occurred during the later part of the Devonian Period, around 370 million years ago. The first was the ability to increase in girth and hence to be able to increase in height—the evolution of arborescence, or the tree habit. It involved two fundamental changes. Until this time all plants only had primary growth; that is, they grew only by cell division at the top of the shoot. The plant stem could not become thicker as the shoot grew taller, so the early plants were limited in height. This is the habit that, for example, weeds and ferns have adopted. Secondary growth around the stem circumference solves this problem. The ability to increase its girth allows the plant to grow continually taller. This technique has proved spectacularly successful for the plant kingdom. But the dawn of arborescence involved not just the trick of secondary growth but also the ability to strengthen the cells, which are predominantly made of cellulose. This involved the development of lignin which, as we saw earlier, is a complex polymer with carbon rings. Lignified cells are much stronger, and provided the rigidity that allowed trees to grow tall. What we know as wood is the growth of secondary xylem tissue, with cell walls composed of around 70 per cent cellulose and

30 per cent lignin. This tough material is also more resistant to decay than plants with cell walls of simple cellulose. So the development of woody plants meant there was greater potential for the build-up of fuel on the ground from dead plant material.

The second development occurred in some groups of plants by the end of the Devonian. Until then, all plants relied on spores to reproduce. The portion of a vascular land plant above ground is known as the sporophyte. We see a modern example of a sporophyte in the form of a fern. Such plants produce spores, which can often be seen in clusters on the undersurface. The spores are formed in fours or tetrads, and each individual spore has half the plant's complement of chromosomes (it is *haploid*). At some stage the spores are released and fall to the damp soil surface. Here they germinate into the next generation of the plant, the gametophyte, an almost invisible stage that grows within the soil and has male and female organs. The male organ releases sperm into the soil that swims to fertilize the female organ on another gametophyte. This then produces another generation of the plant that has the full complement of chromosomes (it is *diploid*), and grows above the soil surface. This type of reproduction requires moist soil, so the plants are limited to damper environments. Some plants found a way to overcome this problem of environmental limitation by developing a vegetative reproduction strategy, spreading by developing clones from upper or underground portions of the parent. This may mean that what appear to be many plants on the surface may in fact be a single plant beneath. In addition, each one of the underground portions could give rise to a new plant. Vegetative reproduction reduces genetic mixing, but gives a survival advantage in the event of catastrophic physical attack. And it allows some expansion beyond damp environments.

At the end of the Devonian, around 370 million years ago, some plants evolved another reproductive strategy. This was the development of the seed habit: In this strategy the sporophyte plant produces two kinds of gametophyte. A female gametophyte produces the egg and is retained by the parent plant in an ovule. The male gametophitic part is similar to a spore, but now called a pollen grain, and produces the male organ and sperm. The pollen reaches the ovule, mainly by wind in early plants but later also by insects, and a pollen tube grows and penetrates the ovule, allowing the fertilization of the egg by the sperm. The fertilized ovule becomes a seed and gains some food reserves from the parent plant before being shed on the ground. This strategy finally released plants from the need for damp conditions and allowed them to spread into much drier environments. Fire now had its kindling.

As fire became an important environmental factor, traits that proffered protection to plants would have been selected for. We have seen that having a clonal habit is useful for a plant in highly disturbed environments. The development of a thick bark in some plants may also have been an advantage. It has been claimed that the development of lignified cell walls played a major role not only in slowing decay and the build-up of fuel, but also in offering some protection from fire.[1] The evolution of a lignin-rich bark that protects the cambium or living layer in the plant may have been a key evolutionary innovation that happened during the Carboniferous Period, 350–300 million years ago.[2]

The Carboniferous Period saw an explosion of plant evolution in terms of diversity, as all these innovations helped plants to spread into an ever-widening number of ecological niches. The trees mainly responsible for the coal we have today were lycopsids (Figure 26). But plant types such as ferns, seed-bearing ferns (pteridosperms), and woody seed-bearing trees, or *gymnosperms*, such as the extinct *Cordaites*, as well as conifers and spore-bearing trees such as *Calamites* (known today only from small herbaceous

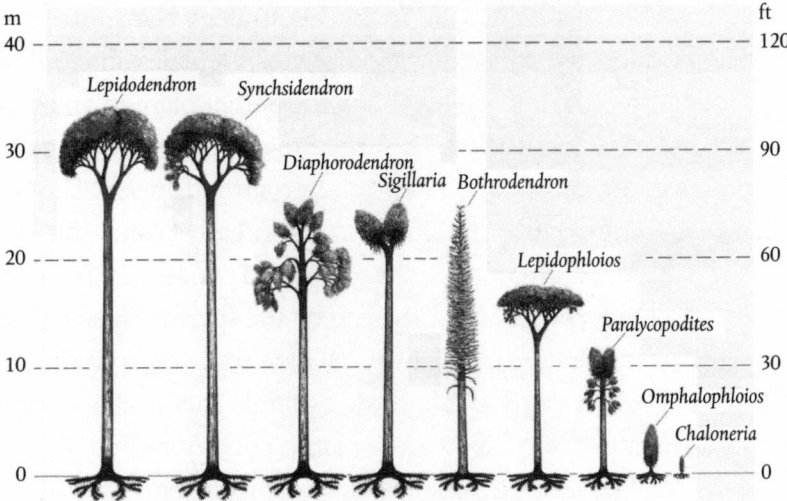

Figure 26. A range of Carboniferous lycopsid trees, the main peat (coal) formers of the period. *Chaloneria* was named after Bill Chaloner.

horsetail plants), all diversified in the Carboniferous. And the evolution of plants with new growth strategies such as lianas may have provided ladder fuels for fires, enabling them to jump from the ground up to the crowns of the trees.[3]

In the southern hemisphere, seed-bearing trees with large deciduous spatulate leaves known as *Glossopteris* evolved, and were particularly important during the Permian Period in the southern supercontinent of Gondwana (Figure 27). Recent research suggests that these plants too developed a fire-resistant bark. Clearly this diversification must have had an impact upon fire systems. In my own research I was able to show that the plants living on riverbanks and on the floodplains were different from those living in the peat-forming mires (swamps and bogs). These ideas have been further developed by several American researchers. Each type of vegetation may have had a different fire regime.

About 250 million years ago, at the close of the Permian, life on Earth underwent a major crisis. This was, as palaeontologist Mike

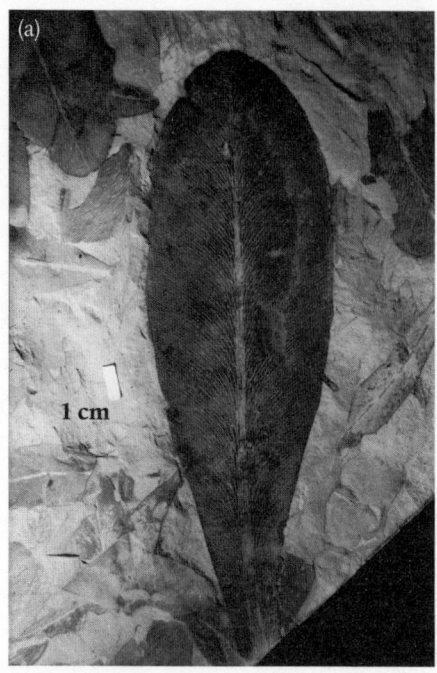

Figure 27. (a) Fossil *Glossopteris* leaf from the Permian (290 myr) of Australia. (b) Reconstruction of a *Glossopteris* forest with an undergrowth of ferns and horsetails.

Benton put it, 'when life nearly died'.[4] The reasons for this mass extinction are complex, but involve a period of significant global warming and volcanic eruptions on a vast scale in Siberia that pumped CO_2 and other noxious gases into the atmosphere. The changes in atmosphere and climate caused many groups of plants to become extinct, including many of the arborescent lycopsids, calamites, Cordaites, pteridosperms, and glossopterids. Life on land in the early part of the succeeding Triassic Period was grim. However, the extinction of some plants led to opportunities for others, and the later Triassic and Jurassic Periods, 250–140 million years ago, saw a major diversification of seed-bearing plants in most habitats across the globe. Groups of plants that came to dominate floras included the conifers, cycads, and an extinct group known as the Bennettitales as well as *Ginkgo*—known today from the single species, *Ginkgo biloba*, or maidenhair tree (Figure 28). Many leaf sizes and shapes evolved, and that may have had an influence on the spread of fire.

By the start of the Cretaceous Period, 140 million years ago, the vegetation was dominated by large swathes of conifers, varying in species across the globe. Lower-lying areas also contained ferns, together with cycads and Bennettitales. But a major change

Figure 28. Living (a) *Ginkgo* and (b) cycad plant types that have survived from the Jurassic Period (145–200 myr).

took place in the vegetation on land through the Cretaceous Period, which lasted from 140 to 66 million years ago: the appearance of flowering plants or *angiosperms*.

The Earth underwent another mass extinction around 66 million years ago, at what used to be called the Cretaceous–Tertiary (K/T) boundary, but is now more properly the Cretaceous–Paleogene (K/P) boundary, and flowering plants reached right up to it. At the boundary, which is marked in many parts of the world by a distinct layer of iridium that is thought to have been caused by an asteroid impact in the Yucatan Peninsula, Mexico, there is a distinct change in the vegetation. Immediately after the catastrophe, ferns dominated for a time. But plants soon became diverse again, and although some species went extinct the overall look of the vegetation was very similar. The mass extinction had a great impact on vertebrate and insect faunas, and the dinosaurs were the most famous casualties.[5] But the turnover of plants was not as great as was once thought.

Over the next 65 million years the modern flora began to appear. The Eocene Epoch, 56–34 million years ago, was critical in the modernization of the flora. It was also at this time that tropical rainforests first evolved and spread across the equator.

The next most important innovation in the plant world was the evolution of grasses. We now know that they evolved and diversified from around 30 million years ago, in the Oligocene Epoch. Grasses produced significant amounts of fuel. However, it was around 7 million years ago that some grasses found another way to photosynthesize using a new biogeochemical pathway, called C4, that allowed them to survive and thrive in much drier environments, forming wide swathes of grasslands. The African savannas evolved at this time. It is only then that a complete, truly modern vegetation evolved on the planet.

Ignition

The second side of our fire fundamentals triangle concerns a source of heat, or a means of ignition. We should consider three natural causes for ignition, with human-originated ignitions being seen only in the past million or so years. When many think of fire in the fossil record they naturally think of volcanoes. After all, there are terms such as the 'Ring of Fire' that describe the volcanoes around the Pacific Ocean. But active volcanism is responsible for starting fires in vegetation only in certain places and times. The occurrence of such volcanic-origin fires across the world is far outweighed by other sources of ignition. A second source is sparks from rock falls. While these do occur, they are even less common than a volcanic source.

By far the most common and widespread cause of natural ignitions is lightning strike. There is a tendency to think that

Figure 29. Lightning is the main cause of natural wildfires.

lightning is only associated with thunderstorms, and hence rain. But various types of lightning occur without rain. Two of the most common forms are cloud-to-cloud lightning and cloud-to-ground lightning (Figure 29). Obviously lightning that occurs over the sea is of little consequence to our story—it is the cloud-to-ground lightning that is our main concern. With the advent of satellites it has been possible to monitor the worldwide occurrence of lightning on a day-to-day basis. What is surprising is the sheer number of lightning strikes across the Earth—it has been estimated that there are some 8 million every day. Far more wildfires are started by lightning strikes than by humans. Even where there are fires caused by human action, additional fires may arise from lightning strike. The key is the presence of sufficient and dry fuel. Under such conditions a fire may spread, whatever the ignition cause. What surprised me were the Yellowstone National Park fires of 1988. I had always assumed that these were started by human activity—campfires or barbecues that got out of control. In fact humans caused only nine of the fires, while 42 were caused by lightning.[6]

How, though, can we determine the causes of fires deep in geological time, using the rock record? This is a particularly difficult question to answer. Clearly there were no human-induced fires before the evolution of humans. This means that for over 400 million years wildfires were started by natural phenomena. Volcanic activity leaves evidence in the rock record, so if in a particular region or period of geological time there was no volcanic activity we can rule it out as a cause. What about lightning? While we cannot observe lightning in the past there are occasions when we can see the effects of lightning strikes. The passage of the electrical discharge produces a very high temperature. If lightning hits sand, the temperature is high enough to fuse the grains, and stringers of fused sand can be found in the soil or on rock surfaces. There are several fossil sites where these *fulgurites*, as they are known, have been recorded. In the British Isles, one of the most famous was recorded in Permian rock, capturing a lightning strike that occurred 260 million years ago in the Isle of Arran in Scotland.[7]

Oxygen

The third part of our triangle concerns oxygen. Combustion is reaction with oxygen, so before there was any oxygen in the atmosphere there could not have been any fire. We know that there was no free oxygen on the early Earth. Oxygen only began to build up, slowly at first, in the atmosphere from about 2.3 billion years ago (what is known as the Great Oxidation Event), with the evolution and build-up of cyanobacteria, which had developed the trick of photosynthesis, in which oxygen is released as a by-product. So at what stage did oxygen levels first rise to a point at which fires could not only start but also spread? We may also ask, when did oxygen in the atmosphere rise to the present atmospheric level of 21 per cent? And what would happen if the oxygen level increased above the modern atmospheric level?

On this last point, why should a higher level than today be a problem? There are two simple experiments that can be done to demonstrate this. The first involves a typical fire in a grate. If bellows are used to introduce oxygen to a smouldering fire, it flames, sometimes quite violently. This action is simply making the normal concentration of oxygen more readily available. Bill Chaloner showed me an interesting and dramatic experiment that I continued to use in my own lectures until 'health and safety' measures made this difficult to undertake. We filled a tube with pure oxygen and placed a cover over it. We then lit a cigarette and let it smoulder. When the cover was removed and the glowing cigarette dropped into the tube, the cigarette exploded in flames! This dramatic experiment demonstrates an important point: if the atmosphere was at any time enriched in oxygen, there would have been more fires, and perhaps hotter and larger than today.

How could the geological record of oxygen be investigated when it could not be directly measured, and we knew of no proxies, unlike the situation for the other gas of particular interest, CO_2?

From the 1970s geologists had attempted to develop geochemical models to calculate the composition of the atmosphere through time, and the Yale geologist Robert Berner produced perhaps the most widely cited models. Bob's first breakthrough had occurred in the late 1980s in a paper with Don Canfield, one of his PhD students, who went on to become a leading worker on the oxygenation of the oceans.[8] Bob modelled both atmospheric CO_2 and oxygen through time.[9] There was much interest, however, in the past 500 million years, during which life developed rapidly on land. These models took account of the long-term carbon cycle—the inputs and outputs of carbon in the system—by taking CO_2 in and out of the atmosphere. One of the keys was the development of photosynthesis by plants—by essentially taking on board microbes that used the trick, in the form of chloroplasts. In photosynthesis, CO_2, water, and solar energy are used to produce carbohydrates, with the release of oxygen. On the death of the organism, the decay of the carbohydrate generates CO_2 again, using up oxygen—the reverse reaction. This can be expressed by the reversible chemical equation:

$$CO_2 + H_2O \leftrightarrow CH_2O + O_2.$$

But if instead the carbon is buried, that reverse reaction doesn't get to occur, so the net result is that oxygen builds up in the atmosphere.

Bob needed to consider the flux of carbon through time. The calculations are complicated, but suffice to say that Berner was able to produce a graph of atmospheric oxygen through time. The resulting curve quickly became adopted by many researchers, but it was of course a model based on various calculations and assumptions. Bob said to me that no sooner had he published one curve, it was time to rethink and recalculate, and he produced a very different curve in his later papers[10] (Figure 30).

Several features of his atmospheric oxygen curves proved exciting and also controversial. His early curves indicated that oxygen levels remained below 15 per cent until around 420 million years ago, and then began to rise. This was as one would expect, as it covers the time when photosynthetic land plants first evolved and spread across the landscape. Then he calculated a rise in oxygen through the early Devonian to around modern levels, with a drop in the middle part of the Devonian before a continuing rise through the later Devonian and Carboniferous, peaking at around 30 per cent in the Permian before dropping back. Oxygen levels appear to have crashed at the end of the Permian, which correlates with the mass extinction event, before rising again back to modern levels over only the last few million years. Particularly exciting were the high levels of oxygen in the atmosphere indicated in the Carboniferous and Permian Periods, well above modern levels. The possibility that oxygen levels were much higher than today was supported by the evolution of giant arthropods, especially the 2-metre-long *Arthropleura* in the later Carboniferous (Figure 31), and giant dragonfly-like insects with wingspans of several tens of centimetres. Such giant arthropods, it was argued, could only survive if atmospheric oxygen levels

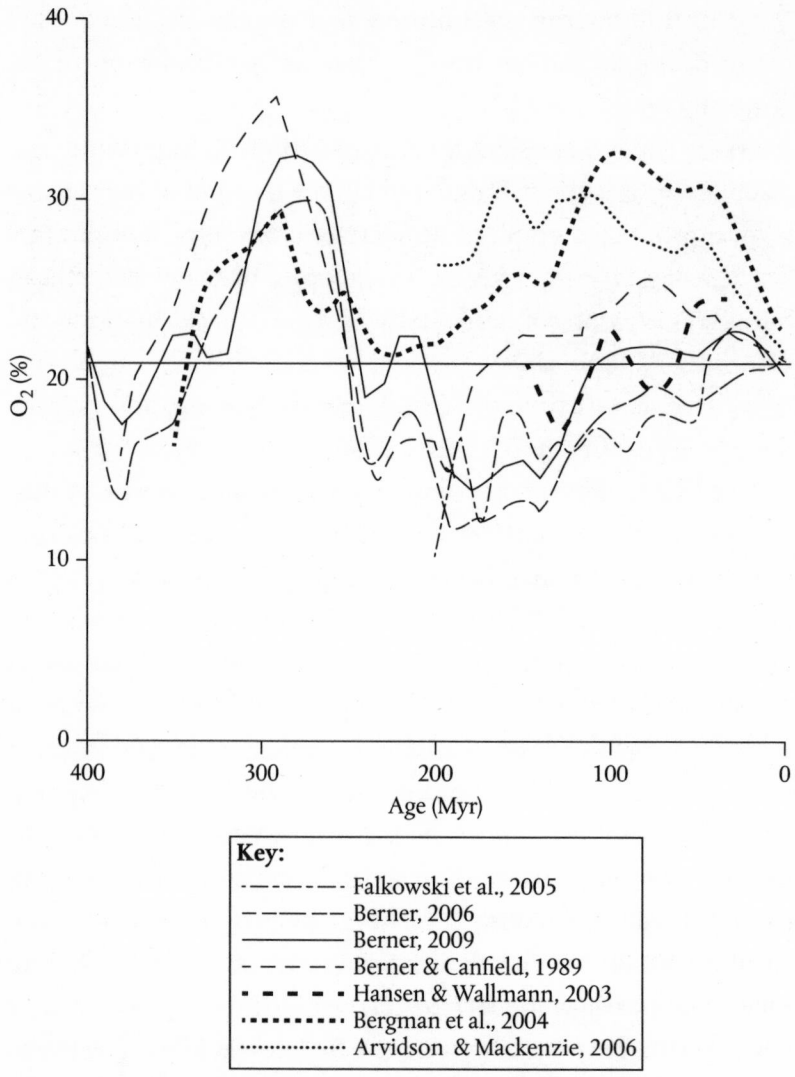

Figure 30. Models of atmospheric oxygen showing a diversity of calculated models. The present atmospheric level is 21%.

Figure 31. Artist's reconstruction of the 2-metre-long *Arthropleura* in the undergrowth of a Carboniferous forest that is beginning to burn. The atmospheric oxygen levels at this time are thought to have been above the modern level.

were high, because these animals used holes in their 'skins' to help the diffusion of oxygen, and a higher concentration of oxygen allows for more efficient diffusion, so the animal was able to function more efficiently and grow much bigger.

Bob Berner was not the only geochemist to model atmospheric oxygen. Tim Lenton, Andy Watson, and their colleague Noan Bergman from the University of East Anglia undertook a very different type of modelling. Although they also found a high level of oxygen in the Carboniferous and Permian, their analysis also showed high levels of oxygen in the Cretaceous, around 100 million years ago. Recent research suggests that these changes in atmospheric oxygen may also have had an impact upon climate.[11] How might wildfire have been affected by changes in atmospheric oxygen, and indeed could wildfires have played any role in the regulation of atmospheric oxygen through time?

We have already seen that the occurrence of charcoal in the fossil record, as evidence for the occurrence of wildfire, could provide a baseline for the atmospheric oxygen level—but what exactly was it? And could we determine the effect of higher atmospheric oxygen levels? Detailed experiments were needed.

The first significant experiments were undertaken by Andy Watson at the University of Reading under the direction of James Lovelock, now best known for his Gaia hypothesis. In these experiments, Andy studied the burning behaviour of a variety of materials, in atmospheres with different proportions of oxygen, looking at the probability of ignition and the rate of flame spread.[12] As well as varying the oxygen content of the atmosphere, he also tried changing the moisture levels in the material being burned. Andy concluded that plants will not burn at levels of atmospheric oxygen below 16 per cent; at 18 per cent fires are suppressed; and as oxygen levels are raised above 21 per cent, even much wetter plant material can burn. At oxygen levels above 30–5 per cent, even water-saturated plants would burn, so it may be difficult for a fire to be extinguished.

Berner and his colleagues tried further experiments, using a range of natural plant materials including peat moss, fern roots, wood, and leaves of the modern conifer *Araucaria*.[13] Their results broadly agreed with those of Watson, though they disputed that completely saturated plants would burn, even in high oxygen. They also showed that conifer needles and wood behaved differently when it came to the start and spread of fire, with a lower oxygen level needed for leaves than for wood. My own interest lay in whether the rise in oxygen concentration had an impact on the temperature of the fire. The equipment used for the oxygen concentration experiments did not have any way of measuring the temperatures. But our work on charcoal reflectance might help, as we had already observed a relationship between charcoal reflectance and temperature of formation. Bob gave me all the charcoal residues from their experiments to see if I could obtain any data, and the results suggest the temperatures of fires do indeed increase with the oxygen concentration.

By the beginning of the new millennium there was increasing interest in the impact of fire in the geological past, and the relationship between fire and the fluctuation of atmospheric oxygen. Charcoal was being widely discovered in rocks of the late Triassic and early Jurassic, around 200 million years ago, that suggested at least a high-enough level of oxygen in the atmosphere for plants to burn.[14] But according to Berner's model, oxygen levels should have been too low at that time to support fire. So where was the error—Berner's model, or the oxygen levels calculated to support fire? Further experiments were carried out in low-oxygen environments. Early results indicated that a more realistic lower limit of oxygen concentration in the atmosphere for fire to start was 15 per cent, challenging the level of 10–12 per cent for the Mesozoic given by the Berner model. Subsequent experiments exploring how oxygen concentration would affect the spread of a fire showed that at levels of oxygen below 18.5 per cent, fire activity would have been greatly suppressed compared to the modern day, and entirely switched off below 16 per cent. Between 18.5 per cent and 22 per cent, fire activity increases rapidly, but above 23 per cent there appears to be only a marginal change in fire spread.[15]

Berner produced revised models, adding new calculations, but these produced even lower oxygen levels for the Mesozoic, including for the Cretaceous, for which his earlier models, like those of Watson and his colleagues, had shown high oxygen. All the charcoal evidence that I was able to gather indicated that oxygen levels across the Mesozoic must have been high.

We still had no way of directly measuring atmospheric oxygen, nor could we find any proxy for its measurement. In the meantime, I had begun to put together a database of all records of fossil charcoal. As I had been able to demonstrate that inertinite found within coal was really fossil charcoal, this gave us another window into fire history. Working with my former student Ian Glasspool, then at the Field Museum in Chicago, we explored the distribution of charcoal in coal over time, and the results proved intriguing (Figure 32).

We had noted that the average figure for the charcoal content in modern peats from across the world was only 4 per cent or so. From the database of fossil charcoal in coal, we found that the figures for the Carboniferous and Permian were very high, usually

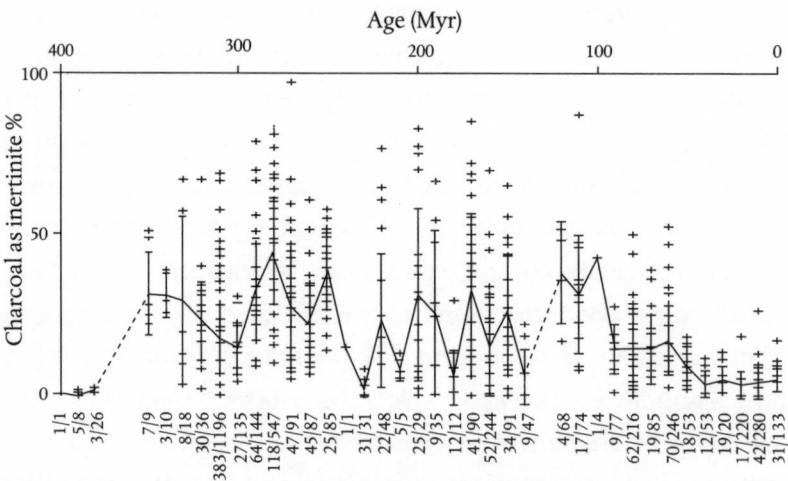

Figure 32. The distribution of charcoal in coal is shown from which the calculated atmospheric oxygen was derived with the line showing the mean of points. The number of data points for each 10-million-year interval is indicated by numbers beneath with a range of points shown and the vertical lines show the standard deviation of the data.

more than 20 per cent but often up to 70 per cent charcoal in coal (this is why coal is often dirty to handle).[16] This was, as all the biogeochemical models suggested, a time of high atmospheric oxygen. We were quite familiar with the large numbers of charcoal bands found within British Carboniferous coals.

As we have seen, experiments had shown that environments of high oxygen would allow wetter plants to burn. As oxygen levels increased in the Permian and Carboniferous, there may have been more frequent fires in the peat-forming systems, resulting in the large number of charcoal bands preserved in these peats, now fossilized into coal.

Another feature evident in the plot of the proportion of charcoal in coal is that there are high levels of charcoal not only in late Paleozoic coals but also in Cretaceous coals, with much lower levels in coals after around 55 million years ago. So it's possible to use the knowledge gained from these data and from the burning experiments to make a proxy for atmospheric oxygen. We would expect no charcoal (inertinite) in coal if the oxygen level was below 15 per cent. We might expect a figure of around 4–7 per cent if oxygen levels were the same as today, at 21 per cent. If we accept the idea that if the oxygen levels were 30 to 35 per cent then fires would be so frequent that fuel would not accumulate, we could use this as an upper figure (obviously the relationship is not linear). The charcoal data can be recalculated and converted to form an atmospheric oxygen curve (Figure 33).[17]

The curve confirms a high level of oxygen in the late Paleozoic (the Carboniferous and Permian Periods, 300–250 million years ago), dropping dramatically at the mass extinction at the end of the Permian. The level fluctuates through the Jurassic (250–140 million years ago), and then rises to the levels of the late Paleozoic again in the Cretaceous. From the end of the Cretaceous, oxygen levels show a steady fall, reaching the modern level by at least 40 million years ago, after which it has remained relatively constant.

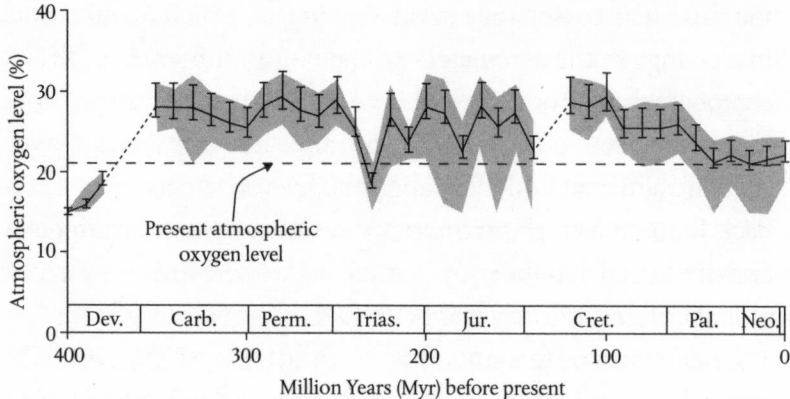

Figure 33. Calculated atmospheric oxygen curve derived from the charcoal-in-coal (inertinite) database shown in Figure 32. The modern atmospheric oxygen level is 21 per cent, and the area of uncertainty is shown in grey.

Feedbacks

The significance of fire triangles is to draw out three factors, and indicate that they are interlinked. They cannot reflect the complexity of the interactions. In reality, aspects such as climate, vegetation, and atmosphere composition form part of an integrated Earth system which also includes the oceans. Changes in one element have knock-on effects on other aspects of the Earth system, which may be positive or negative, and which then in turn may have a return influence or feedback on the first. If the feedback is positive, the changes in the two elements build up; if negative, the changes are damped down and the system tends back to its original, stable state. Fire is a player in these feedbacks, and we must allow for feedback loops when we try to understand the variations in fire in the past and try to predict changes in fire regimes in the future. If, for example, there is a long-term drop in rainfall, it becomes drier, and this may mean that there are more fires. It may also lead to a change in the vegetation, which in turn results in a change in the associated animal communities. But burial of charcoal arising from the fires, in the form of stable carbon, would reduce the CO_2 in the atmosphere, and that would cause global cooling, which would then tend to reduce fires in a negative feedback loop. In fact, the artificial production of charcoal (biochar) and its burial has been suggested as a mechanism to reduce global CO_2 levels today and help hold back global warming.

Understanding feedbacks has become increasingly critical in a complex world, and one of the first to examine feedback in fire systems was Lee Kump, of Pennsylvania State University, of whom more later. Bob Berner was also much engaged with feedback loops, and used 'systems diagrams' to help visualize feedbacks

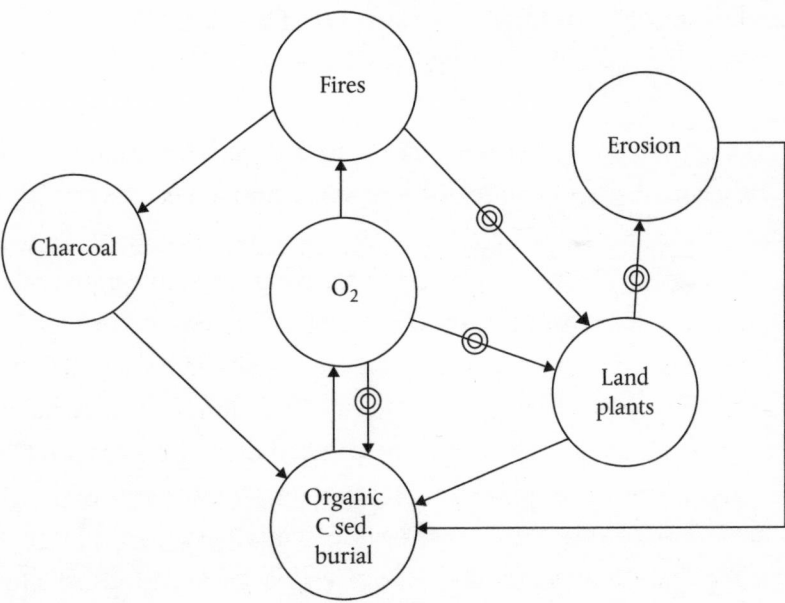

Figure 34. Earth systems model showing the relationship between fire and oxygen. Arrows show a positive feedback, and arrows with circles indicate a negative feedback. An even number of arrows indicate an overall positive feedback.

and calculate whether an overall effect is positive, negative, or neutral with respect to the Earth system. One of his classic diagrams illustrates the relationships between fire, charcoal, vegetation, atmospheric oxygen, erosion, and carbon burial (Figure 34). A good example of how these interrelationships work in a complex feedback loop is to consider what happens as oxygen levels increase. If oxygen increases, then fires increase. If there is more fire then there is a decrease in land vegetation. If there is less land vegetation then there is more erosion. If there is more erosion then there is more carbon burial, and that in turn causes an increase in atmospheric oxygen. Essentially a single change can have extensive knock-on effects. Some of these feedbacks have a short duration, but others have a much more lasting impact on Earth system processes. We will look at some of these feedbacks as they become important in our story of fire through time.

4

The rise, fall, and rise of fire

When I started my doctoral research in October 1973 I never imagined that I would spend so much of my career thinking about fire. I had not considered fire as an agent of change on Earth, or that charcoal deposits may preserve its long history on the planet. I had never thought of fire as a preservational mechanism for fossil plants, producing charcoal that would show their anatomy so that they could be identified, and help us to piece together the vegetation that must have clothed the land millions of years ago. In all my years of collecting fossils as a child and student I had never found, or at least noticed, any fossil charcoal.

I had wanted to look at the ecology of the plants that were found during the Carboniferous, 300 million years ago. The natural approach was to look at the large fossil plants that could easily be found in rocks such as the Coal Measures that are often found scattered on old coal tips. But many smaller plant fragments are also preserved in the rocks. I started a programme of dissolving the rocks in acids and obtaining residues of the fossil plants that remained. The rocks are made up of minerals that dissolve in different acids from the plant fossils, which are made of organic material. It was hard work, and I spent many hours a day picking through the plant fragment residues, which were about the size of tea leaves, trying to identify what the fragments represented. Incredibly, at that time, few researchers had tried to look at plant fossils in this way. I soon noticed a large number of fragments that looked like charcoal, and examined these with an SEM.

Under the SEM the astonishing detail in the charcoalified leaves was revealed (BW Plate 6). The small needle-like leaves had two beautifully preserved rows of stomata. But what kind of plant did they come from? I took the material to Bill Chaloner, who was one of the world's authorities on the lycopods, one of the most common plants found in the coal measures. Only a few species of this group, such as *Lycopodium*, the club moss, are found living today. After much discussion, and work collecting more material (one leaf per day!), there was sufficient material, including two leafy shoots, to become convinced that the fossil plant was a new kind of conifer. This made the discovery doubly significant, as it represented the oldest known conifer, and it was preserved as fossil charcoal. We speculated that the conifer may have been living outside the immediate area, perhaps in upland terrain, and was subjected to wildfire. The charcoal would have been washed down by rivers and deposited on the low-lying floodplains. Much to my surprise, my paper was accepted by the prestigious science journal *Nature*, and appeared just as I began the second year of my doctorate (it was to be another 30 years before I had another *Nature* paper!). And so began my lifelong fascination with fire, and what it can capture of the past.

The earliest fires

To look for evidence of the earliest fires, we needed to search rocks of the Devonian Period (419–360 million years ago), when land plants became sufficiently established to provide kindling. We had some charcoal specimens from this time, and new material was collected from Germany, but the records of Devonian charcoal were scarce and it was not clear if this was because it had not been recognized or collected, or whether there was a real paucity.

The world's leading authority on the earliest land flora is Dianne Edwards from the University of Cardiff. Dianne had puzzled for some time about the preservation of some of her fossil plant specimens dating even earlier, back to the Silurian Period (around 420 million years ago), and we had corresponded about the possibility that some of them could have been charcoal. Working with one of my former students, Ian Glasspool, they combined SEM imaging of these tiny plants with the reflectance techniques that we had developed at Royal Holloway, to demonstrate that some of their plant specimens were indeed preserved as charcoal. Here was evidence of some of the very first wildfires on the planet.[1]

This was followed by research on the earliest charcoal from the Devonian.[2] What was becoming clear was that there were examples of fire in the earliest period when plants first spread on to land in the late Silurian and early Devonian, from 420 to 395 million years ago. Fires would not have been large or widespread at this time, as the distribution of plants would have been patchy and limited to areas near water, as they had only developed a free-sporing reproductive strategy. The build-up of fuel would have been slow, as the plants could only grow using primary growth. There were no trees at that time.

One intriguing plant (or possibly a giant alga or lichen) that sometimes appears preserved as charcoal from this time is *Prototaxites* (Figure 35).[3] This appears to have been a very large plant and has been reconstructed as a tall upright pole. Surely with such a shape the plant may have acted as a lightning rod, so its preservation as charcoal would not be a surprise. These early charcoalified plants also showed that oxygen levels in the atmosphere

Figure 35. Reconstruction of the Devonian (410 myr) possible alga or lichen *Prototaxites*. The tall, thin shape of the 'plant' might have attracted lightning strikes.

were high enough by the end of the Silurian and into the early Devonian for wildfire to occur. Other records of charcoal were restricted to the later Devonian. What about the middle part of the Devonian, from 390 to 380 million years ago? We know there was an extensive flora by then, and indeed the Devonian was a time when plants were increasing in size.

Ian Glasspool and I puzzled about the Devonian charcoal problem for some time, and we went out of our way to try to track down material. We then started to look at the possible reasons for the scarcity of evidence of wildfire in the mid-Devonian. One of the obvious culprits was the atmospheric oxygen content. If this had dropped below 17 per cent then fires would not have started or spread, and sure enough Bob Berner's models all proposed a dip in atmospheric oxygen at this time.

Fire then appears to have returned to the world in the later Devonian. American researchers had previously found charcoal in rocks from north-central Pennsylvania dating from the late Devonian, which at the time had been the oldest known. The late Devonian, around 360 million years ago, was a time when trees and hence the first forests evolved. We had always assumed that the first forests coincided with the first forest fires, but this does not seem to have been the case. Trees and forests are found before the first extensive charcoal deposits. Most of the charcoal from the late Devonian rocks of Pennsylvania was from a ground-scrambling fern-like plant called *Rhacophyton*, and not from the tree known as *Callixylon/Archaeopteris* (the trunk and the leaves had been found separately and each given a name—a common occurrence for fossil plants).[4] We had, therefore, evidence of surface fires but not of crown fires, or at least not of fires within the forests.

So when did the first forest fires occur? In Belgium and Germany there was increasing evidence of charcoal more commonly preserved in terrestrial and marine sediments from the latest Devonian. Belgian scientists had described a number of charcoalified specimens. Some of these were indeed of *Callixylon*. But samples were still rare. There was, though, another possible approach to the problem of the timing of the first widespread fires.

We saw earlier that charcoal can be lofted by the wind and travel very long distances. It may also be waterborne, travelling

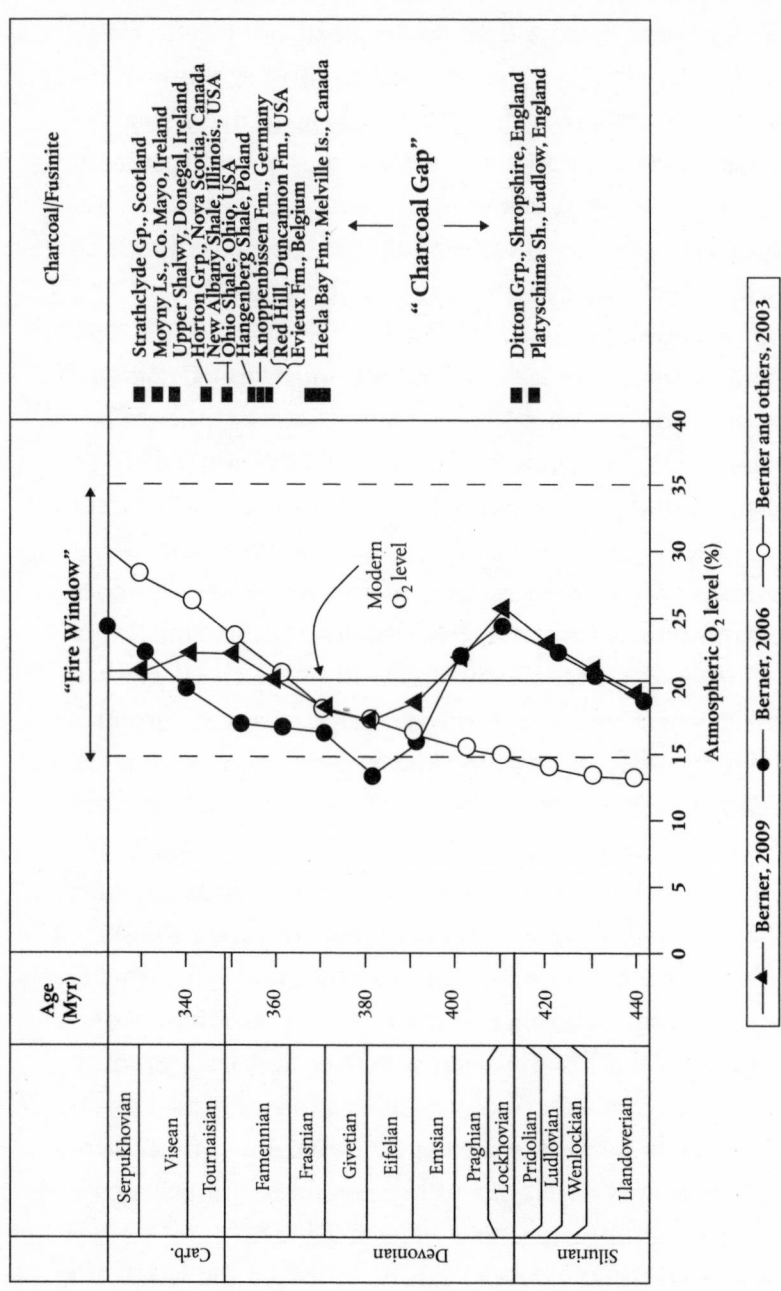

Figure 36. The rise of fire in the late Devonian as recorded by fossil charcoal records, according to a number of models. The rise of fire may relate to the rise of atmospheric oxygen at the time. The fire window is the range of oxygen levels that will allow sustainable fires.

Plate 1. Fires in California as seen from a satellite on a single day in October 2007, with smoke plumes billowing out over the Pacific Ocean. Position of active fire shown coming from a single area.

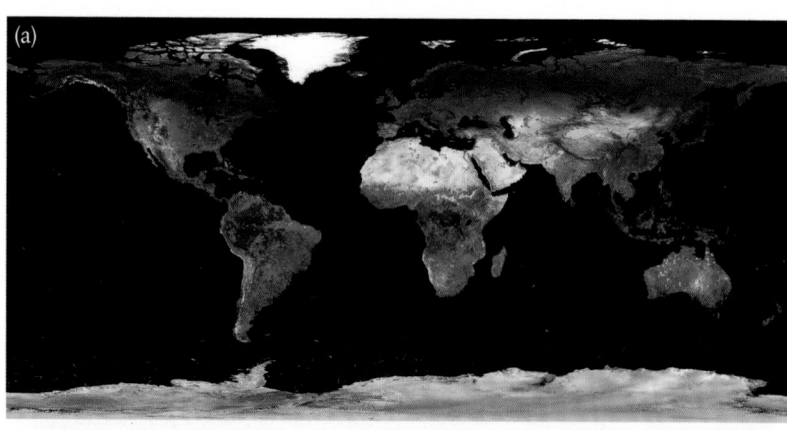

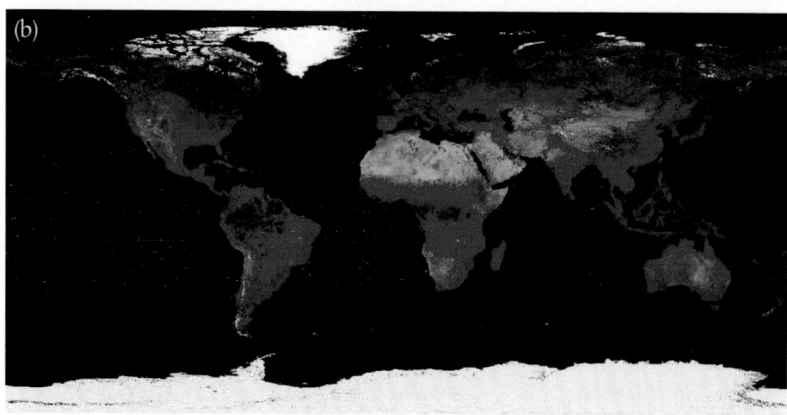

Plate 2. Fire across the world as recorded through satellite observations. (a) Active fires seen during 1–10 January 2013; (b) map showing all fires collated through the year for 2016.

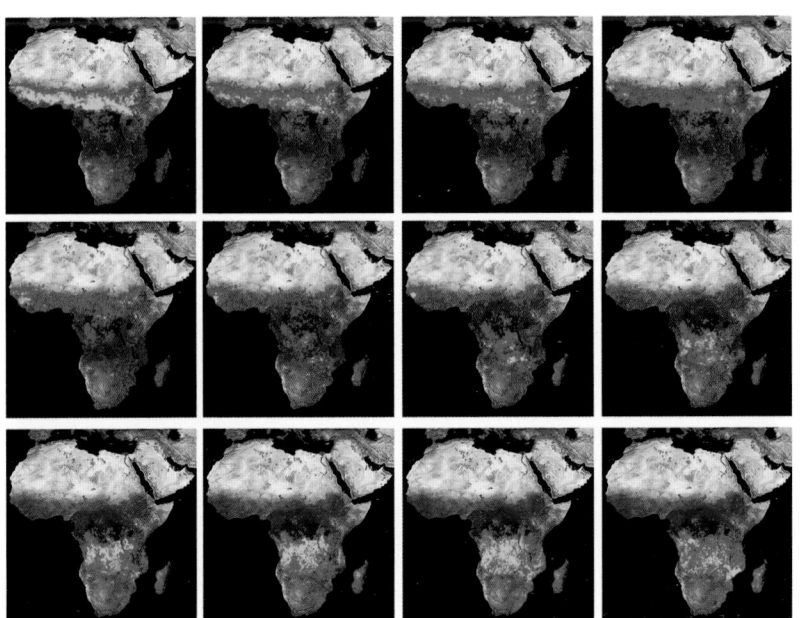

Plate 3. Fire in Africa, as recorded through satellite observations, showing strikingly different patterns at different times of year. Each image shows cumulative data for fires across a ten-day period each month, over eight months in 2005, from January (top left) to August (bottom right).

Plate 4. The 2015 North Fire in southern California jumped the freeway to destroy houses and put people at risk.

Plate 5. Thick dark smoke from fire in a coniferous forest.

Plate 6. Deer taking refuge in a river from a wildfire in North America.

Plate 7. Jack pine/black spruce forest fire in boreal Canada. Note the surface fire has spread via ladder fuels to the crowns of the trees.

Plate 8. Fire in dry forest in south-east Australia, a type of vegetation known as sclerophyll, with *Eucalyptus* trees. The predominantly surface fire has spread to the crown of one tree.

Plate 9. Charcoal in brown coal from the Miocene (20 myr), brown coal (lignite) deposits near Cologne, Germany. The charcoal pieces are black and shiny 1 cm cubes.

Plate 10. Black charcoal and brown uncharred megaspores from Lower Carboniferous sediments (325 myr) from Berwickshire, Scotland with well-preserved charcoalified stems.

Plate 11. Sandstone-filled trunks of arborescent lycophyte, Upper Carboniferous (300 myr) at Joggins, Nova Scotia.

Plate 12. Reconstruction of a tetrapod at Joggins, Nova Scotia of Carboniferous (300 myr) age, where the trees may have been hollowed out by one fire and may have acted as a den or hide during another wildfire event.

Plate 13. Reconstruction of the Cretaceous high-fire world, showing dinosaurs (*Struthiomimus*) fleeing a fire.

Plate 14. Peat burning fires in Indonesia.

along river systems to the sea. Charcoal can be identified in marine sediments, even as very small particles, so any widespread increase in fire, especially if there had been little before, could be picked up in marine sediments. American coal petrologists had begun to look at the composition of marine sediment in the eastern USA from the latest Devonian Period to the earliest Carboniferous, around 365–355 million years ago. They found that there was an increase in fossil charcoal, from the late Devonian, at just the time when oxygen levels in the atmosphere had begun to rise, according to most models.[5] Unfortunately our charcoal-in-coal proxy for atmospheric oxygen is difficult to use for this time period, as there are very few coals of this age anywhere in the world. More data from different parts of the world were needed in order to show that the rise in fire indicated from the end of the Devonian was a truly global phenomenon. Sure enough, the pattern was found to be widespread. All the evidence now suggests that the very end of the Devonian, around 350 million years ago, was a period when extensive wildfires first occurred in the fossil record (Figure 36).[6] We were now seeing a world with fire where the atmospheric levels of oxygen were beginning to approach those of the modern day.

The Carboniferous

As we have seen, charcoal becomes abundant in rocks of early Carboniferous age, such as the 345-million-year-old rocks I described from Donegal in Ireland. Further rich charcoal deposits were later discovered in Ireland by my research student Howard Falcon-Lang, on the north Mayo coast, in sediments deposited in a river estuary associated with many fossil fish. These rocks were the first in which an environmental catastrophe resulting from a wildfire could be documented. The fire had caused sediment and charcoal to flood into the estuary, killing the fish.[7]

Charcoal was also discovered in many of the rock sequences of Scotland, including some early Carboniferous deposits, from around 330 million years ago, in the Borders region on the east coast, as well as north of the Firth of Forth (BW Plate 7). SEM studies were able to show that a range of plants were preserved as charcoal (BW Plate 5, Colour Plate 10), and the diversity of occurrence indicated that fires were frequent in a range of environments at this time. They may have been started either by lightning or by the volcanism that was abundant in the area.

The rise in fire on the planet was dramatic. From a period in the mid-Devonian in which there was little or no fire, by the Carboniferous, between 350 and 320 million years ago, fire had become ubiquitous. This was just the time for which Bob Berner had calculated a significant rise in atmospheric oxygen. Evidently, there was still much to do on unravelling Carboniferous fires and their impacts on the vegetation and wildlife of the time.

In 1984, Stan Wood, an amateur palaeontologist who became famous for his fossil collection, had discovered one of the oldest known tetrapods (four-legged land animals) from a quarry at East Kirkton near Bathgate, not far west of Edinburgh. Stan had originally discovered the fossils in a farmer's wall surrounding a football pitch. He was refereeing a match and at half time went to look at the limestone blocks making up the dry-stone wall. He not only found the oldest harvestman spider (one with a small body and very long legs) but also a complete small tetrapod.[8] Stan did two things that were typical of him. He went to see the farmer and bought the limestone that made up the wall. And he looked into where the stone had come from, discovered that the quarry was disused, and bought the mineral rights from the local council. He arranged with the Royal Scottish Museum to keep half the quarry for his economic activity of selling fossils, but gave the other half for scientific research. An international team of more than 50 researchers was assembled to excavate and study the quarry over the next five years. Research on the material is still going on today. The animals and plants were preserved in what had been a lake that was supplied with hot water from a nearby volcanic source. However, we found large quantities of charcoal in the lower part of the section, especially in the levels associated with the vertebrate fossils, including the now famous 'Lizzie the lizard'.[9] Fire was an important element of the ecosystem, particularly in the lower part. It's possible that the fires may have driven animals to take refuge in the waters of the lake, but as it was toxic, they died there, and became preserved in the sediments. Many of these volcanically active areas of Scotland, from around 340 million years ago, frequently experienced wildfire, whether from volcanic activity or lightning strikes, and their rocks contain abundant fossil charcoal.

There are other indications, too, of the response of Carboniferous animals in regions swept by wildfires. Joggins, in Nova Scotia, Canada, tells us much about life in the Carboniferous. Many layers of rock were discovered with fossil trees (Colour Plate 11). Charles Lyell visited Joggins with the famous Canadian scientist William Dawson in the 1850s, and they discovered some of the world's earliest tetrapods within hollow fossil tree trunks.[10] This preservation has fascinated geologists ever since. When I visited Joggins myself, I noticed a lot of charcoal within the sedimentary sequence and even within some of the tree trunks. Large *Cordaites* trees were also associated with the charcoal, and it is thought that these grew in upland areas that were swept by fire. Several environments where fire was an important part of the ecosystem are represented at Joggins, and that has allowed us to speculate on the origin of some traits that evolved in plants that would have helped them cope with fire, such as the development of thick bark layers and the shedding of leaves and branches from tree trunks that would prevent fire migrating from the surface up the trunk to the crowns of the trees.

The tree trunks were in some cases hollowed out by fire, so it is possible that fire played a role in their preservation and fossilization. Fire may also have been the reason for the fossil vertebrates associated with charcoal, a number of which have been found within the hollows. The hollowed-out tree trunks may have acted as refuges from the fires for small animals that were frequently a part of the ecosystem, and some of them may have been killed there as the fire passed (Colour Plate 12). The fire-hollowed trunks

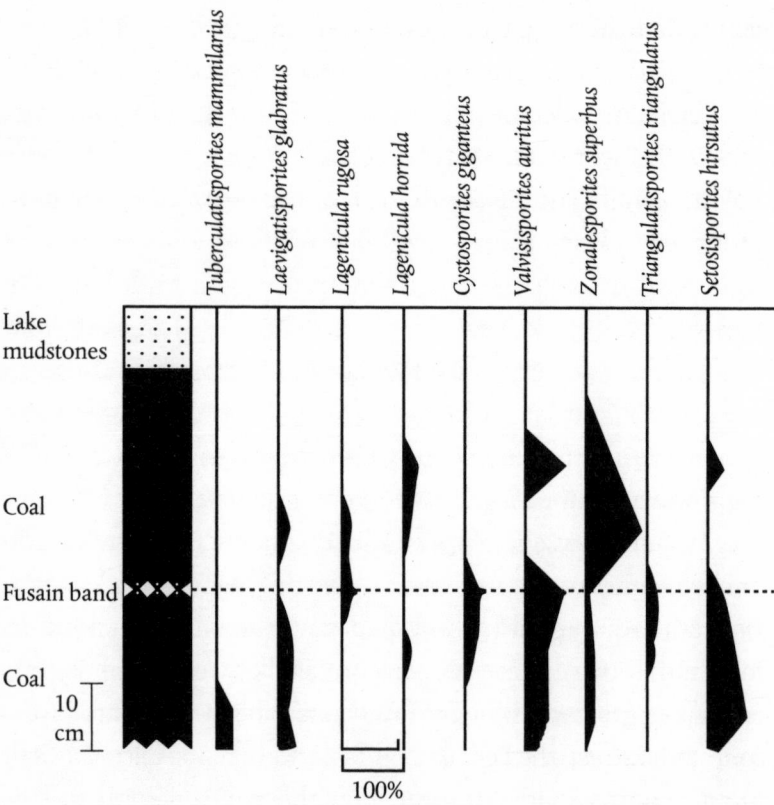

Figure 37. Vegetational change through an Upper Carboniferous (310 myr) coal seam from Yorkshire showing the charcoal (fusain) horizon.

mass, we would expect to find coals of Permian age there too. One of the goals of the doomed Scott expedition to the South Pole was to try to find fossil plants and Permian coal, and in this they succeeded. The specimens of *Glossopteris* that they found helped to confirm the existence of the palaeocontinent Gondwana. Scott kept the specimens with him until the end. They are now housed in the Natural History Museum in London.

In the 1980s the coal-fired power stations in the British Isles almost exclusively used British Carboniferous coals. Following various miners' strikes, the power companies began to diversify the coals that they burned. For the first time Chinese coal was imported to the UK. I myself spent some time studying Chinese coals. However, following the turmoil of the collapse of the Soviet Union, there was a determined effort by the Russians to develop their coal for export. One of the mining areas was the Permian Kuznetsk (or Kuzbass) Basin, in Siberia, which would have been in the northern hemisphere in the Permian. Several power companies had begun to import this coal to burn in the UK. When the coal was burned, it became clear that its combustion behaviour was different from the Carboniferous coals found in Britain. One possible reason for this was the high charcoal content of the coal. Was this a common feature in these coals? And if there was evidence of frequent fires, might it be possible to calculate the time gap between them—what is called the fire return interval? If the atmosphere was enriched in oxygen then the frequency of fires may have been enhanced.

Obtaining permission to undertake research on Siberian coals involved complex negotiations and sensitivity, both economic and political. We were keen to go not just with an interpreter and other trade representatives, but a coal geologist as well, so we could explain more clearly what we required for our research. Natalia Pronina, a coal geologist from Moscow State University who spoke very good English, proved a valuable colleague for this research. My research student, Vicky Hudspith, and I flew to Moscow and then on to Siberia. It was a long flight to Novokuznetsk, some 200 miles to the east of Novosibirsk on the eastern side of the Ural Mountains, just north of Kazakhstan. In the middle of the first night, I was woken by shouting and a loud banging at my door. All I could make out was the word 'passport'. Was it a police raid? It turned out that the hotel was on fire. We all got out safely, but it took time to put out the fire, and we had to be lodged elsewhere for the rest of our stay. A strange reminder of the continuing power of fire, even as we explored its deep history.

The coals in the Kuznetsk Basin were quite different to those with which I was familiar. Many were over 10 metres thick, as opposed to the typical 1–2-metre-thick coals in the UK, and there were several coals exposed in what is one of the world's largest open pit mines.

In spite of the unfortunate hotel incident, we were able to go ahead with fieldwork in the mine, and it was successful. We were able to show that the coals did indeed have high charcoal content, and that fires had been common within the peat-forming areas and the surroundings during the Permian. We went on to work out the fire return interval. To do this, we first needed to calculate the original thickness of the peat, by working out how much it had been compacted through burial. From the position of the continents and from climate indicators, including the occurrence of growth rings in permineralized and petrified wood, we could deduce that the peats formed in a temperate climate. Our knowledge of the rate of accumulation of modern peat in temperate climes provided an estimate for our fossil peat. So we could now estimate the time taken for a particular thickness of peat to form, and, knowing the number of charcoal horizons, we could work out the fire return interval in the peat-forming system. The results indicated that fire return intervals were shorter in the Permian than for the modern day in such settings.[12] This supports the idea that fire systems were affected by the high levels of oxygen in the atmosphere at the time. Charcoal was soon found in localities across the world from this time, and forest fires appear to have been common (Figure 39).

Life in a high-fire world is hard to imagine. Today we have both fires that start naturally and those lit by humans, but humans also put out fires, so their natural frequency and extent are difficult to judge. But we can expect that in a high-fire world, fires were probably more widespread across different climate regimes; they were probably large, sometimes intense, and more frequent than today. As a result, vegetation growth may have been more restricted, and as for animals, a variety of these would have been

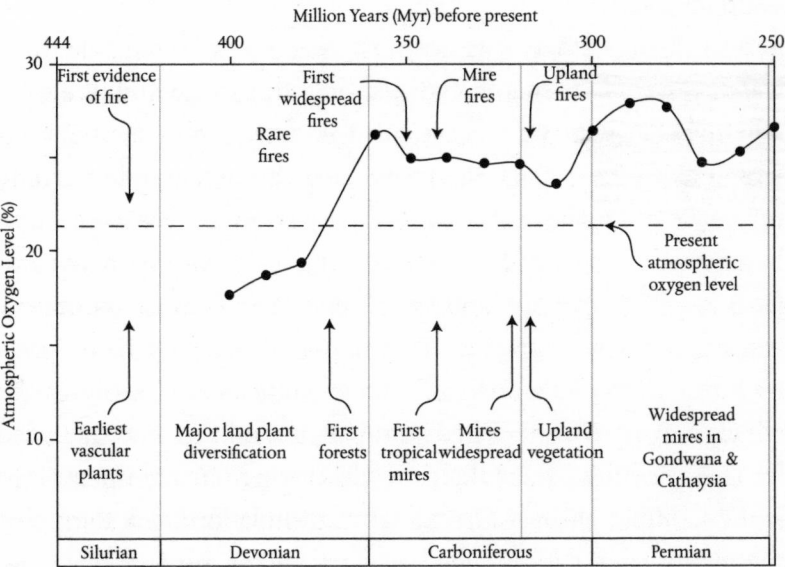

Figure 39. The evolution of fire regimes during the late Paleozoic (350–250 myr) in relation to atmospheric oxygen.

affected, including those of the air, as smoke might have been much more of a problem even than today.

At the end of the Permian came the biggest mass extinction event in Earth history. Although its causes are still under debate, a combination of the release of methane from the oceans following the end of a major ice age, and major volcanic eruptions in Siberia, would have led to an increase in CO_2 and poisonous gases. Temperatures rose significantly and pollutants contaminated both land and sea.

What happened to fire systems at that time? Investigating charcoal in the rock record can help to answer this. Sediments in China cover the period from the latest Permian up to the early Triassic. Chinese coal petrologists were able to show that there was significant charcoal in the coals right up until the end of the Permian.[13] Fires were clearly an important part of the ecosystem throughout the period. Indeed, data from coals in eastern Yunnan show increasing fire activity towards the end of the period. At this stage, it is not possible to say whether the fires were a result of the major changes happening at the boundary, or whether they were part of the cause of the changes. I have often wondered at the impact the smoke from the large number of fires may have had upon the ice caps of the period. We know that carbon particles deposited across snow and ice fields darken the surface, causing the snow and ice to absorb more solar radiation. Large areas of blackened vegetation would also have had a similar effect. This could have led to a small warming of the Earth, which may in turn have had some impact on the melting of the ice sheets of the time, which would further darken the surface and encourage more warming, as a positive feedback loop. But this is only speculation.

The events at the Permian–Triassic boundary, around 250 million years ago, were game changing. Some 95 per cent of species became extinct. From a fire point of view, this had two important consequences. Plants that had adapted to a high-fire world died out, and models of atmospheric oxygen all suggest a rapid fall in oxygen at the boundary. The Triassic Period that followed was a very different world.[14]

5

Fire, flowers, and dinosaurs

The Mesozoic Era is the geological interval comprising the Triassic, Jurassic, and Cretaceous Periods, and it is best known for the rise and fall of the dinosaurs. The Mesozoic began around 250 million years ago and continued to around 66 million years ago—a not inconsiderable chunk of geological time, and framed by mass extinctions at its beginning and end. Fifty years ago there were very few published papers on fire in deep time, but the most important one, which I've touched on before, was 'Forest fire in the Mesozoic', by Tom Harris of the University of Reading.[1] Tom was an important scientist, one of the leading palaeobotanists in the world. Energetic and passionate about his fossil plants, he was a scientist with broad interests, and given to experimentation and lateral thinking. The evidence that Tom used in his paper on fires in the Mesozoic was limited to only a couple of charcoal occurrences in these rocks.

The Triassic

The Permian Period ended with the biggest known mass extinction in Earth history, when life was almost wiped out. Whole ecosystems collapsed. So what would the world have looked like at the start of the Triassic?

Among whole groups of plants that had become extinct were the giant club mosses that had been the major coal-forming plants of the late Paleozoic, and the glossopterids that had dominated southern continental vegetation. In the first few million years after the extinctions, plant diversity appears to have been low, but some new plants became prominent, including the pole-like spore-bearing lycopod called *Pleuromeia*, and the scrambling seed-plant called *Dicroidium*, which had fern-like foliage (Figure 40).

Figure 40. Reconstruction of Late Triassic vegetation showing *Dicroidium* in the foreground, along with lycopsids, cycads, tree ferns, *Araucaria*, *Cylomeia*, ginkgophytes, and conifers.

The first 10 million years of the Triassic are thought to have been a time of ecosystem recovery. According to Berner's model, the Triassic started with very low levels of oxygen in the atmosphere.[2] Researchers had noticed that there were no coals found at the beginning of the Triassic, and this interval was called the 'coal gap'.[3] The problem, therefore, was that charcoal in coal could not be used as a proxy for atmospheric oxygen for this time interval. And there were very few, if any, records of charcoal in early Triassic sediments. It was unclear whether this was because there were no fires because of low oxygen levels, or because the vegetation was patchy so large fires did not occur, or simply because no one had been looking for charcoal in sediments of that age. One significant problem is that there are very few places where early Triassic sediments occur, and where they do, they are mainly in marine environments, where charcoal is less likely to be noticed. But work on early Triassic fossil soils concluded that they were indeed formed in a low-oxygen atmosphere.[4]

Although charcoal has not been found in the earliest Triassic sediments, what researchers did find were a series of deposits of mid and late Triassic age (247–201 million years ago) containing charcoal.[5] This charcoal was mainly of gymnosperm wood, so large trees had reappeared by the mid-Triassic. Our understanding of fire systems at this time is rather limited, as the charcoal recovered was mainly of wood, with a limited number of species represented. But it seems that, 10 million years on, the planet had finally recovered from the devastating events at the Permian–Triassic boundary. By this time, many new plants had begun to diversify, not only cycads but a range of seed-bearing plants, as well as ferns and horsetails. In the animal world, the first diversification of the dinosaurs had begun.

The end of the Triassic and the boundary to the Jurassic, 202–199 million years ago, is a particularly interesting time, and formed the main subject of Tom Harris's 1958 paper on fire. He described limestone topography in Cnap Twt, in South Wales, with fissures filled with sediments from the very end of the Triassic, 208–201 million years ago. The main plant preserved as charcoal was a conifer called *Cheirolepis*. Harris imagined fires in upland conifer forests across South Wales, with the sediment, charcoal, and unburned plants becoming washed into the fissures.[6] Late Triassic fires were also thought to be responsible for the formation of some vertebrate *bone beds*—sediment layers in which many bones have accumulated—perhaps resulting from erosion and subsequent deposition following a catastrophic wildfire.[7]

Harris also recorded charcoal from across the Triassic–Jurassic boundary in East Greenland, and in recent years Greenland has once again become the focus of research into this time period. A large number of fossil plants can be collected from the rocks crossing the Triassic–Jurassic boundary. This allows us not only to track the changing vegetation through this time interval, but also to look at changes in atmospheric CO_2 and in global temperatures. As we have seen, the analysis of the stomata of fossil plants can give information on CO_2 levels. There were major changes in both at this time, perhaps partly caused by the breakup of the supercontinent of Pangea, in which all major landmasses on Earth had briefly come together by the end of the Permian. The end of the Triassic saw one of the five mass extinctions in Earth history, though there is continuing debate on the causes. There was a major change in climate that appears to have been driven by a sudden release of CO_2 into the atmosphere. This in turn had an impact on the vegetation. The flora changed across the boundary, with a switch in vegetation from a broad leaf dominated assemblage to one in which needle leaves predominated. This coincided with an increase in fire activity just above the boundary, based on the charcoal record.[8] The question is, why? The main leaf shapes had changed, and laboratory experiments showed that broad leaves burn differently from needle leaves, with needle leaves being the more flammable. The warming associated with increased CO_2 in the atmosphere, coupled with an increase in lightning activity, may have led to an increase in wildfires. Another study of coals spanning the Triassic–Jurassic boundary in Sweden also revealed a major increase in fire activity across the boundary, again using the distribution of charcoal. In this rather different setting the scientists found that the vegetation changed from coniferous forest to shrubby vegetation, and using our charcoal reflectance/temperature method they were able to show not only changes in fire temperatures, but also a shift from high-temperature crown fires to lower-temperature surface fires.[9]

The Jurassic

Berner's later models indicated very low oxygen levels throughout the Jurassic (from 200 to 145 million years ago).[10] Clearly this could not be the case, as charcoal from fires is commonly found throughout this interval of time (Figure 41). Our knowledge of Jurassic fires is still, however, rather limited. We need to return to that paper by Tom Harris. After his work on Greenland, Tom spent much of the rest of his career describing the flora from the Middle Jurassic rocks of the Scalby Formation in Yorkshire, making it one of the most famous fossil plant localities in the world (Figure 42). He noted the occurrence of charcoal in the rocks north of Scarborough. Most spectacular are the sandstone beds full of leaves of *Ginkgo*, and he recorded wood charcoal fragments as well as occasional charcoalified ferns. I collected material myself from the locality and investigated the charcoal with my

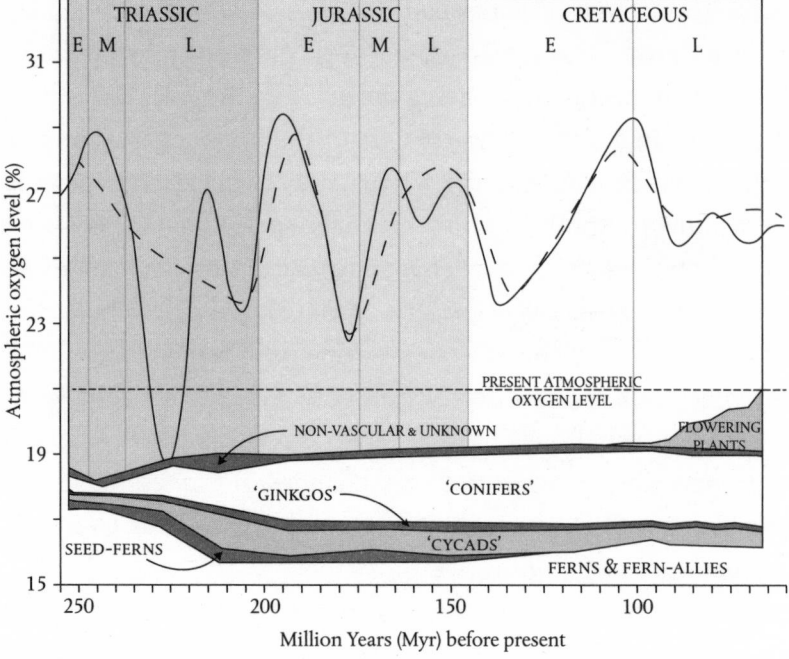

Figure 41. Changes through the Mesozoic (250–65 myr) in vegetation and in atmospheric oxygen as calculated from the charcoal-in-coal proxy. The solid line and dashed line represent mean values calculated in two ways, using a 5 million- and 10 million-year time slice respectively.

students. New research by Mick Cope showed that the floras affected by wildfires included conifers, cycads, Bennettitales, and Ginkgoales.[11] The world at this time was, it seems, dominated by seed-bearing plants, with a ground cover of ferns and horsetails. There were still no flowers. The dinosaurs that roamed across this landscape included giant herbivorous sauropods and carnivores such as *Allosaurus*.

There do not seem to be many occurrences of charcoal in these Jurassic rocks. I had collected at many localities but believed that it might be rather restricted. Others have looked for charcoal throughout the rock sequence, and investigated the relationships

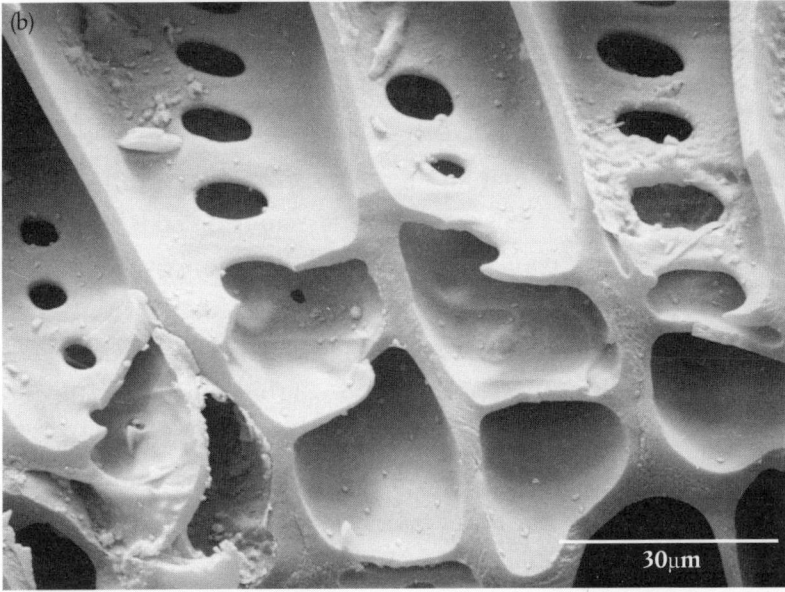

Figure 42. (a) Charcoal from the fluvial sandstones of the Middle Jurassic (170 myr) Scalby Formation, Scarborough Yorkshire; and (b) as seen under the SEM.

between the sediments, plants, and climate. They found that charcoal was restricted to particularly dry periods—i.e. that fire was climate limited and not oxygen limited, and the climate state oscillated through that time period. Britain was at that time north of the equator, on the northern side of Pangaea. The southern side of Pangaea consisted of the ancient and vast continent of Gondwana, and research in Argentina, India, and Antarctica all similarly demonstrates evidence of wildfire at several levels in the middle part of the Jurassic. But we still have much to learn about fires in these environments, not only about what was being burned, but also on how these fires affected the animals.

It had occurred to me that there must be some evidence of charcoal in the many sequences in the North Sea area. After all, many of the sands that acted as reservoirs for oil were similar to those sediments exposed on the Yorkshire coast that had contained charcoal, including the Brent Sandstone Group. This is perhaps best known for the oil it contains, 'Brent Crude', the main price indicator for North Sea oil. I suggested to one of my students, Tim Jones, that he should look for charcoal in the drill cores. What surprised us was just how many horizons with charcoal he found.[12] It was well established that some of these sands were eroded from upland areas to the north, but the shedding of sand into the basin had always been associated with tectonics—i.e. tectonic uplift followed by erosion of the rising land. But perhaps they were pulses of sediment and charcoal shed off the hills as a result of post-fire erosion? If that was the case, then some of these sedimentary packages could be very widespread and individual units could be correlated. I was reminded of the regular occurrence of sediment and charcoal recorded following modern fires in the Rocky Mountains. Such sediment pulses could occur on a regular basis. These studies of North Sea cores showed that charcoal was more abundant in some units than others, so fire activity seems to have fluctuated. And some charcoal layers occurred towards the end of the Jurassic.

In the late Jurassic, during the time interval known as the Kimmeridgian (157–152 million years ago), charcoal has been reported from western and central Europe, but only a few charcoal fragments have been recorded, and scientists have argued that at this time oxygen levels were on the lower end of the spectrum that could still allow fires.[13] Our charcoal-in-coal data also

Figure 43. Purbeck fossil forest from the latest Jurassic (140 myr) of Lulworth, England. A petrified conifer tree that has been removed from the rock.

suggest a lower atmospheric oxygen level at this time, so it was not just climate that controlled the occurrence of wildfires.

At the very end of the Jurassic, around 150 million years ago, there is again some evidence for wildfire. In this case many of the fires appeared to occur in coniferous forests. The best examples come from the southern coast of England, known as the Jurassic coast, near the Isle of Portland, and from the well-known fossil forest at Purbeck (Figure 43).[14] These conifer forests were spread along the coast in a climate not unlike parts of the Middle East today. From time to time the forests burned, and charcoal is preserved in the soil.[15] The trees were studied by Jane Francis, who also showed that there were conifer forests across Antarctica at this time.

The early Cretaceous

By the beginning of the Cretaceous (140 million years ago) the supercontinent of Pangaea had begun to break up, with a number of modern continents being recognizable (Figure 44). The rocks of the Weald, in southern England, are non-marine rocks dating back to this time. These rocks were being deposited as the North Atlantic Ocean was just beginning to form, and they are world famous for their fossil plants and their dinosaur fossils, which were some of the first dinosaurs to be described in the early and middle part of the nineteenth century. The plants are mostly preserved as compressed fossils, mainly of leaves. Famous palaeobotanists such as Sir Alfred Seward and Marie Stopes had described fossils from the deposits, and Percy Allen from the University of Reading conducted a monumental study of these continental sediments.[16]

Ken Alvin, of Imperial College London, was well known for his work on fossil ferns from the Cretaceous in Belgium, and had done

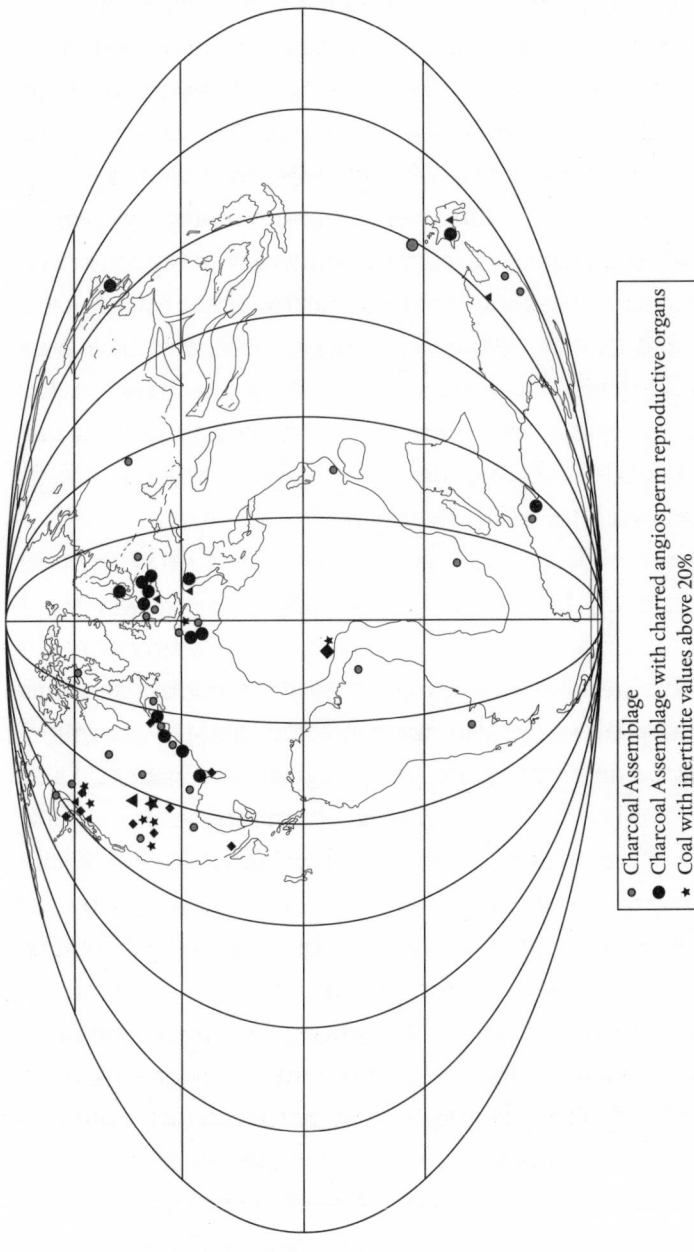

Figure 44. The distribution space of Cretaceous charcoal (140–66 myr) on a palaeogeographic map.

pioneering work using the SEM. These two interests came together when he discovered charcoalified ferns in the Wealden rocks of the Isle of Wight, off the southern coast of England.[17] These sediments were again of early Cretaceous age (145–125 million years ago), and our understanding of the environment of that time expanded rapidly in the early 1970s. Percy Allen had originally built up a picture of deltaic sediments across the Wealden area, a model that became widely used. But at a geological meeting in 1975, Percy astounded the whole audience by showing that his deltaic model was probably incorrect and that the sediments were more likely to have been deposited by rivers.[18] Such is the nature of science. The change in interpretation of the environment had many implications, not least for the study of the Wealden climate, and also for the ecology of the plant fossils found in the deposits.

I had been collecting charcoal from the Isle of Wight over many years, and when I moved to Royal Holloway we undertook a more systematic study of the occurrence of charcoal in the Weald, and especially from the Isle of Wight sections. We found that the charcoal occurred at many intervals, and the plants represented were different at different horizons (BW Plate 8). There were two main communities that had been affected by wildfire. The first was coastal fern 'prairies', and the other was coniferous forest.[19] The questions remained: was this increase in burning more widespread? And how might the fires have been affected by both oxygen concentration and climate change?

Evidence of similar charcoal deposits, and hence of wildfires, soon came from other localities. In Nova Scotia, borehole cores taken from subsurface basins with early Cretaceous sediments turned out to have horizons full of charcoal, which contained conifers and ferns reminiscent of the Wealden rocks of southern England.[20] This was clearly a high-fire world. Drilled boreholes form a marvellous resource for scientists studying sediments and fossils, as they neatly capture the layering of rocks to a considerable depth, which are not always exposed on the surface. One of the oldest known pines was also found in the same locality, supporting the idea that this group evolved during a high-fire interval in Earth history.[21]

Visiting a quarry in Belgium with rocks of the same Lower Cretaceous age, I looked for examples of charcoal to compare with our material from the Isle of Wight. As always the most exciting finds occur in fading light as you are leaving the site. Within a sandstone channel, I saw a lump of charcoalified forest floor that had obviously been washed into a river. It proved to be full of charcoalified plants and even some insects, and their preservation was quite spectacular (BW Plate 9). The occurrence of charcoal in such rapidly deposited river sediments was probably a result of post-fire erosion.

The appearance of flowers

By far the most important event in the evolution of plants during this period was the origin and spread of flowering plants, or angiosperms. What would strike any time traveller (apart from the dinosaurs) is that in the early Cretaceous greens and browns dominated the landscape, but by the end of the Cretaceous the scenery would have been much more varied, with some views a riot of colour.

The origin of angiosperms was called by Charles Darwin 'an abominable mystery'.[22] So, from the middle of the nineteenth century to the early part of the twentieth century, the search was on for older and older plant fossils that could show when this important evolutionary development occurred. In 1912, Marie Stopes claimed that she had found angiosperm woods from the

Lower Cretaceous rocks of southern England.[23] There was no question that the fossils she found were of angiosperm woods. But unfortunately the specimens she described were from museum collections, and the provenance of the specimens, and hence their age, was uncertain.

During much of the twentieth century there were claims and counter claims of the oldest angiosperms, and even some suggestions that they had evolved more than 200 million years earlier, in the Triassic. Not only was the discovery of the first angiosperms important, but so was establishing how quickly they diversified and spread, to clothe the world in colour. There are still several candidates for the oldest angiosperm, from China and also even from the Wealden rocks of southern England.[24] But there are other ways to seek the origin of angiosperms: we can look for pollen that is distinctive to them. Spores and pollen of the Wealden had been investigated through the 1960s and 1970s. As we might expect, in the earliest Wealden rocks angiosperm pollen was rare, and it became gradually more common and diverse as you progressed upwards through the layers, and thus to younger times.[25]

In the USA too, especially in the eastern parts, there were several early Cretaceous sequences that had yielded plant fossils, and these, and the pollen found, helped to provide a much more detailed picture of the evolution and diversification of the angiosperms. At certain levels, one of the plants found in numbers on its own (a 'monotypic assemblage') was *Sassafras*.[26] This was considered to be abundant as a result of a disturbed environment, and sure enough, some of the levels immediately below the beds in which *Sassafras* occurred were found to contain plenty of fossil charcoal.

We have already seen the importance of the discovery of charcoalified flowers in Sweden. These plants were from a sequence that was from near the end of the Cretaceous, around 70 million years ago. What about other deposits? A wider search for fossil plant assemblages that contained flowers ensued, and many of them turned out to be preserved as charcoal. Discovering them is not an easy task. These early flowers were often small, only around a millimetre or so in length. Only rarely are small charcoal particles found or recognized on rock surfaces. It is usually necessary to remove many kilogrammes of rock or sediment and process it back in the laboratory to look for small preserved flowers.

Over the next 20 years new floras were discovered and described, many of which contained charcoalified flowers (BW Plate 10).[27] Some deposits were very rich and contained other

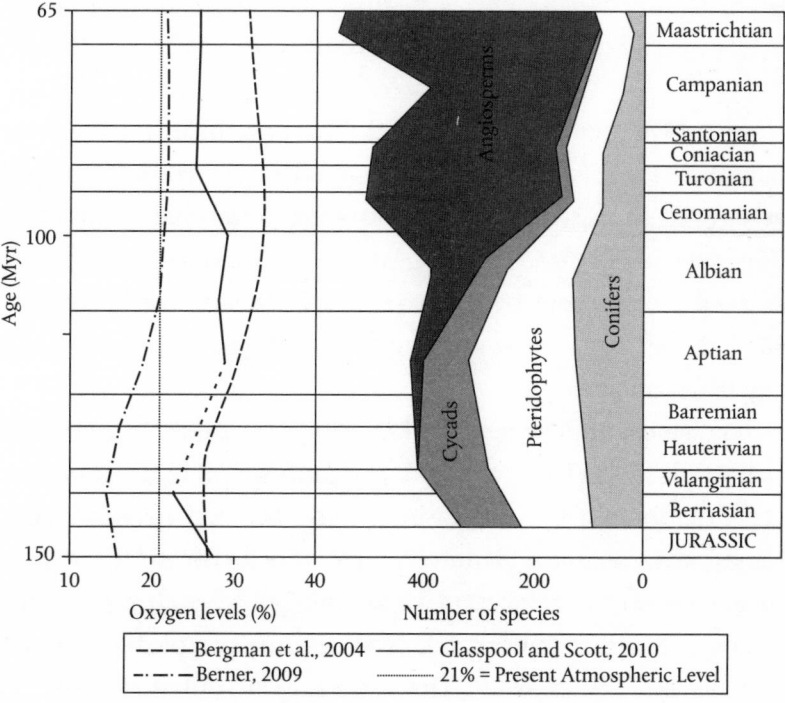

Figure 45. There are major changes in vegetation through the Cretaceous, during a high oxygen and high fire world.

plants; some also contained fossil woods. Charcoalified floras containing flowers soon turned up across the globe, even as far away as Antarctica, which was green during the Cretaceous despite being at the pole.[28]

Evidently, the Cretaceous was rich in fossil flowers, indicating how quickly flowering plants must have spread and diversified (Figure 45). It was also a high-fire world. Early angiosperms appear to have been predominantly weedy plants, such as *Sassafras*, which did well in disturbed environments. Fire is, of course, a major disturbance factor. It seems likely, then, that fire may have aided the evolution and spread of the earliest flowering plants.[29]

Fire and dinosaurs

How did the high-fire world of the Cretaceous affect dinosaur communities? Could some of the dinosaur bone beds have formed as a result of post-fire erosion, flooding, and rapid deposition? This question first arose during studies on the Isle of Wight. One of the main dinosaur beds, the *Hypsilophodon* bed, contained fossil charcoal, but there were simply not enough data to come to any firm conclusions.

The Dinosaur Provincial Park in Alberta, Canada, is one of the richest places on Earth to find dinosaur fossils (Figure 46). Many of the dinosaurs are on display in the Tyrell Museum in Drumheller, Canada. No charcoal had ever been recorded from these sediments, yet if the Cretaceous was a high-fire world it surely must be there. Imagine my surprise when I came out of the museum and climbed the steps to a viewing platform, only to see several horizons with fossil charcoal. Other beds with dinosaurs, in Texas and France, have since also proved to contain charcoal.

Figure 46. Dinosaur-bearing Upper Cretaceous sediments (90 myr), from Dinosaur Provincial Park, Alberta, Canada.

New research on dinosaur bone beds is at last taking into account the roles of fire and post-fire flooding events in the formation of at least some of them.[30] Fire in the late Cretaceous was a significant influence, and must be taken into account when we reconstruct ecosystems. Some artists now put fire and dinosaurs in the same scene (Colour Plate 13).

The evolution of fire traits

Fire may have had impacts on evolution too. One aspect that we considered at another period of high fire in the fossil record, the late Paleozoic, was the evolution of traits that allow plants to cope with fire. The Permian extinction essentially 'reset' the evolution of such traits. We know that a number of plant groups in the modern world have traits to cope with or even take advantage of fire.[31] When, then, did these traits evolve?

Until about 20 years ago, unravelling the occurrence of traits would have been restricted to a study of fossils. But advances in molecular biology have led to the development of what is called molecular phylogeny. This technique uses the differences in DNA code between living forms, and the approximate rate of mutation, as a molecular clock, allowing us to extrapolate back to establish the relationships between groups, and estimate when they split off from a common ancestor. We can use this method to trace the origin of traits in particular branches. Such an analysis of the family Pinaceae, consisting of pines and their relatives, shows that traits such as a fire-resistant bark had their origin in the high-fire world of the Cretaceous.[32] The Proteaceae, a family of flowering plants that includes *Banksia*, also have many fire traits which can be traced back to the Cretaceous.[33] This conclusion, based on the DNA of modern plants, has been confirmed by fossil evidence, in Cretaceous rocks from Australia, of the earliest proteaceous plants.[34] The earliest pines, too, have been found as charcoal fossils in Canada, directly linking the group with wildfire. The clear indication from two lines of evidence that the fire traits of a number of plants today originated in the Cretaceous gives further strong support to the idea that the Cretaceous was indeed a high-fire world.

Fire may have impacted on ecosystems in other ways. Research on modern wildfires suggests that fire may affect the phosphorus cycle. Phosphorus is a nutrient essential for plant growth, as all gardeners will know—it is an important fertilizer. Lee Kump showed the impact that fire had on the phosphorus cycle.[35] If phosphorus is released by a fire, it has the possibility of being transported to other environments, and affecting plant growth there. If there were very large fires in the Cretaceous, this may have led to a temporary influx of phosphorus into the marine environment. This would result in the rapid and extensive growth of marine algae. The decay of much of this rapid-growing but short-lived vegetation uses up oxygen in the water column, producing ocean anoxia. Such anoxic intervals show up in the rock record in the form of black shales that are rich in organic material, and early research suggests that some of these may show fire markers.

What then would the world have looked like at the end of the Cretaceous, 66 million years ago? Clearly this was a much more modern world as far as vegetation was concerned, with both coniferous and flowering trees across the landscape, but it was still a world dominated by dinosaurs. We can only wonder how they would have managed in this high-fire world, with large swathes of vegetation being regularly burned.

A global conflagration?

The extinction of the dinosaurs is a subject of constant fascination, to scientists as well as the general public. The discovery in 1980 of a thin layer of iridium that could only have come from an asteroid impact in sediments at the Cretaceous–Paleogene (K/P) boundary across the world, pointed to one possible cause (Figure 47).[36] The claim was not without controversy, and even though the Chicxulub crater in Mexico was subsequently identified as the site of impact, there remains some disagreement as to whether the impact was the *only* cause of the end Cretaceous mass extinction. The period also saw major volcanism, in the form of huge outpourings of lava that produced the flat terrain now known as the Deccan Traps in India. Such volcanism would also have had profound effects on the atmosphere and altered the climate. But

Figure 47. The Cretaceous–Paleocene (formerly K/T) boundary, United States (66 myr). The boundary layer (white) contains melt spherules from the asteroid impact. Fires, however, were a frequent occurrence during the Cretaceous and evidence does not support a global wildfire after the impact.

the relevant point to our story is that soon after the impact hypothesis was published, some researchers proposed that, following the impact, there had been a global wildfire.[37] This hypothesis was based on soot found in deep-sea sites in several parts of the world. The idea was appealing and caught on, so that a global conflagration is included in most reconstructions of the Chicxulub impact even today. But is it true?

I had my doubts. There were two types of problems with the claim: the first was our knowledge of how fires start and spread;

and the second was the nature of the evidence itself. As we saw in Chapter 1, fire does not occur naturally in all ecosystems. The possibility of fire depends on three factors—fuel, level of moisture, and ignition. There must be sufficient fuel for a fire to burn and spread. We have seen that fuel is not evenly distributed, so that a fire may run out of fuel. We have also seen that the moisture content of the fuel is critical. If the fuel is too wet, then a fire will not burn—all the energy goes into evaporating the moisture rather than breaking down the cellulose and lignin to provide flammable gases. Some areas are particularly wet, and for them to dry out may take a considerable amount of energy. As we have seen, however, a higher atmospheric level of oxygen in the atmosphere than today may have allowed wetter plants to burn, but even then not all plants would catch fire at the same time.

After igniting, fire spreads, but there are natural barriers such as lakes and rivers, and some plants live in very wet soils near water. It is hard to imagine how fire could start at the same time everywhere. So a worldwide conflagration would seem most unlikely. However, if somehow all the vegetation were ignited at once, the resulting fire would be very hot, and all animals would be killed by such a maelstrom, even those that burrowed into the soil.

There would be other consequences too. In particular, large areas would be stripped of vegetation, leading to extensive post-fire erosion. We should see burned peat surfaces and potentially vast quantities of charcoal from living plants. More than half the charcoal from five K/P sites was found to be from rotting plant debris and not living vegetation. What's more, there is no evidence of significant fires following other meteor impacts.[38]

The evidence arguing for a global wildfire was in the form of the amount of soot in the boundary layers and geochemical markers for combustion, both from marine sites. But we know that fires were frequent in the late Cretaceous, and fire markers, including charcoal, can easily be transported to the oceans. The soot deposit would need to be clearly linked specifically to the impact layer, rather than the result of fires before or immediately after the impact, as dead vegetation dried out and may have been ignited by lightning. The transport, deposition, and preservation can also lead to concentration effects in marine environments.

What about evidence from terrestrial sites? There is a locality in New Mexico, called Sugarite, where the impact beds were found within a coal seam. This was useful for determining the charcoal (inertinite) distribution through the coal—before impact, through the impact layer, and after. Not surprisingly (at least to me), ample evidence of fire both before and after the boundary was found, with no special concentration at the boundary. There was no evidence of a burned peat surface or of post-fire erosion.[39] This was the nearest terrestrial site to the Chicxulub crater associated with the impact, so would presumably have been most affected.[40] Although we had our doubts, more detailed work was needed on the distribution of charcoal at terrestrial impact boundary sites. A series of studies across North America was begun from the south of the United States to across the Canadian border by my colleague Margaret Collinson and I, and our research student Claire Belcher. Large blocks of rock were excavated that spanned the boundary. These were cut, and polished cross-sections prepared, so that we could study the charcoal bands in place in the rocks, knowing exactly where they occurred, with respect to the impact layer.

The work confirmed that charcoal occurred before and after the boundary, and although some was found within the impact layer, it was not of a different character or in any greater concentration.[41] Of course, the impact layer was deposited in a short period of time, but as we have already seen from modern wildfire studies, large quantities of charcoal can accumulate very quickly after a single wildfire event. And again, there were no signs of heated surfaces or of any post-fire erosion deposits.[42]

One reason for the persistence of the idea of a global fire was that early models of the asteroid impact suggested that very high temperatures would have been produced. Surely, the argument went, if temperatures were so high then there must have been significant fires? But the calculated temperatures were later revised downwards as models improved.[43]

What about the evidence from the soot and the geochemical markers for combustion? New research suggested that many of the soot particles were typical of the combustion of fossil fuels.[44] The rocks in which the asteroid had impacted were found to have contained fossil fuel deposits, which could have vaporized during the impact. What's more, geochemical markers that were collected from the terrestrial sites had compositions more typical of fossil fuel combustion than the burning of living vegetation.[45]

So now we seem to have reached a good consensus. Fires occurred repeatedly in the hundreds of years before and after the K/P boundary event. Most scientists agree that some local fires were probably started by the impact, but the temperatures were not high enough, and did not last long enough, to have produced a global conflagration. The appealing myth, though, will probably persist.[46]

6

Fire and the coming of the modern world

What kind of world dawned after the K/P boundary? We know from studies across localities in the USA that there is evidence of frequent wildfires continuing into the earliest Paleogene. But what happened to the atmospheric oxygen level after recovery from the K/P mass extinction—did it remain above modern levels? Were we still in a high-fire world? If there were fires, what is the evidence in the charcoal record, and do we know anything about the vegetation that was burning?

When the charcoal in the coal database was originally compiled, one of the important issues was how we recorded and represented our data. Early to mid-Paleocene Epoch coals (from around 65 to 55 million years ago) are often recorded as 'earliest Tertiary' in coal literature. (The Tertiary was the name we used to use for what we now call the Paleogene and Neogene Periods, stretching from around 65 to 1 million years ago.) However, coals that are nearer to the start of the Eocene Epoch, just older than 55 million years ago, are notoriously difficult to date. This is a problem we have with many coal sequences, as they are deposited on land, and most of the fossils used to give ages are found in marine waters.

Many coals of this age are often simply recorded as coming from the late Paleocene or early Eocene. Where we have good dating information, Paleocene coals all tend to have high inerti-nite (charcoal) contents, well above 19 per cent. By the mid to late Eocene (50–40 million years ago), however, worldwide the char-coal contents are low, around 5 per cent or even less. There must, therefore, have been a fundamental change in the Earth system at this time.

Another problem is the way in which we chose to represent our data and show the calculated oxygen curve. In order to get sufficient data to plot the curves we decided to use 10-million-year bins. This was not a problem for the Paleozoic–Mesozoic transition, covering the great Permian mass extinction, which took place 250 million years ago. The data were placed in two bins, covering 250–260 and 250–240 million years ago, and the change across the boundary could be clearly seen. But the Cretaceous–Paleogene boundary (the transition from the Mesozoic to the Cenozoic, covering the K/P mass extinction) occurs at around 66 million years ago, and the Paleocene–Eocene Epoch boundary is at 56 million years ago—so that the bins covering 70–60 and 60–50 million years ago both encompass major geo-logical boundaries. Changes across them wouldn't be apparent with our system. All the same, an analysis of the undisputed Paleocene data (mainly from the earliest Paleocene) suggests high oxygen, but undisputed mid to late Eocene data suggest that oxygen levels were stable and at around the present atmospheric level of 21 per cent. Indeed, it appears that after around 50 million years ago we were entering a world with the present level of oxy-gen, so that oxygen level was no longer an important influence on wildfire activity.

So what about the Paleocene? I have always been surprised at two facts. First, despite the common occurrence of charcoal in the Paleocene *coals* there were no charcoal records in the litera-ture from Paleocene *sediments*. What's more, there were no records of charcoalified flowers, despite the fact that so many had been described from the Cretaceous. Perhaps no one had actually looked? I set colleagues a challenge while on sabbatical at Yale University: select a number of Paleocene sediment samples, and I bet there will be some charcoal in at least some of them. Sure enough, this proved to be true. Good charcoal turned up in samples from collections at Yale and the Smithsonian Institution in Washington, showing excellent preservation of anatomy. The material still awaits an enthusiastic researcher to follow it up.

The boundary between the Paleocene and Eocene Epochs, 56 million years ago, is defined in a sequence of marine rocks in Egypt. It has always been the practice to use characteristic fossils to help fix the boundary. However, in this case it was decided to try a very different method, involving *isotopes* (different forms of a chemical element that vary in the number of neutrons they possess). This method has only been possible in the past few decades because of advances in our understanding of the chemistry of the oceans and methods to measure particular types of chemical change.

The field of isotope geochemistry has advanced rapidly in recent years. In particular, there has been much interest in what are known as stable isotopes. Radiogenic isotopes undergo radioactive decay and can be used to find the absolute age of rocks. But the proportions of stable isotopes may vary depending on the environment, and may then be incorporated into plant and animal skeletons. That means we can use the proportions of such isotopes within fossil plants and animals to probe the environment of the time in which they lived.

Carbon has three main isotopes, with 6, 7, and 8 neutrons. These isotopes are described according to their atomic masses (the total number of protons and neutrons) as carbon-12, carbon-13, and carbon-14, respectively (or ^{12}C, ^{13}C, and ^{14}C). Carbon-14 is radioactive, with a half-life of only around 70,000 years, and so is useful for dating archaeological artefacts. By far the most common isotopes of carbon are carbon-12 and carbon-13. These isotopes are stable, with carbon-12 being the most common. Animals and plants preferentially incorporate the lighter isotope, carbon-12, into their skeletons. Carbon exists in a cycle through the Earth system, with some becoming buried in the form carbon-12-enriched organic matter, while other carbon is released into the atmosphere via volcanic activity, so the overall carbon isotopic composition of the atmosphere and oceans can change. In other words, the ratio of carbon-12 to carbon-13, $^{12}C/^{13}C$, is not constant. We have learned that through time there have been many changes in this $^{12}C/^{13}C$ ratio, and that at some periods the value moves well away from the norm. This is known as a major carbon *isotope excursion*, and indicates a significant disturbance to the Earth system. As such excursions are potentially global, we can use them to correlate rocks. Such an excursion is observed in rocks near the Paleocene–Eocene boundary, and a good example, in a suite of marine limestones at Dababiya in Egypt, was chosen to define the boundary for the International Geological Time Scale (Appendix).

There was evidently something perturbing the Earth system around the Paleocene–Eocene boundary. But what was it? The advantage of using the carbon isotope excursion was that it could be found in both marine and terrestrial rocks, getting round the problem that the fossils commonly used in correlation in marine sequences are not found on land, and vice versa. For the first time we could compare events happening in the oceans with those happening on land.

Isotopic studies on marine fossil shells and limestones had shown that oxygen isotopes were also changing at this time. The oxygen isotopes concerned are stable isotopes, oxygen-16 and oxygen-18. The ratio of these oxygen isotopes is affected by changes in temperature, and this can be picked up by the records left in shells and ocean sediments.[1] These studies indicated that a rapid, short-lived rise in global temperature occurred at the base of the Eocene. This has been named the Paleocene–Eocene Thermal Maximum, or PETM. During this event, which may have lasted less than 20,000 years, global temperatures are thought to have been raised by between 5°C and 8°C. The onset of this period of global warming, during which temperatures rose by around 5°C, was very fast in geological terms, and may have taken only about 2,000 years.

Research on global isotopic shifts from the PETM onwards has built up a very detailed picture of changing global temperatures. It appears that even in the Eocene there were other, smaller perturbations of the global temperature, but following the brief warm spell we were inexorably into a fall in mean global temperatures, taking us from a 'greenhouse' to an 'icehouse' world.

The key questions remained: why did the PETM occur, and what effect did it have on life and the planetary environment? We live today in a time of rapid climate change. As we grapple with the rise in atmospheric CO_2 caused by the burning of fossil fuels, and the resultant increase in global temperatures, the lessons we might learn from the PETM have become especially pertinent.

One initial culprit for the rapid change in climate was the possible release of methane from the ocean floor. Methane is an even more powerful greenhouse gas than CO_2.[2] Not everyone was happy with this idea, and other calculations suggested that there must have also been a release of carbon from terrestrial environments.[3] Another suggestion was that many of the peatlands were subjected to wildfire, causing a rapid increase in global atmospheric CO_2 levels.[4] Here was another boundary event in which fire was potentially implicated.

I had discussed fire in the Eocene with colleagues over many years, but after more than 30 years of diligent looking they had found virtually no charcoal in any of their deposits—a fact that was a great disappointment to us all. Geologists had studied the vegetation from around the Paleocene–Eocene boundary, but no one knew quite where the boundary was, or whether the carbon isotope excursion could be recognized in those rock sections.

A breakthrough was close to hand, from the south of Britain. At the present time, southern Britain is at a latitude 51°N, but around 55 million years ago was at 40°N, equivalent to the central Mediterranean today. It was also warmer than it is now.

One of the major infrastructure projects of the late twentieth century was the building of the Channel Tunnel between Britain and France. It was clear that the rail link to the tunnel was not sufficient and that a high-speed rail track was needed. We now enjoy sitting back in our seat and watching the world pass by at a blur. But how many now speeding along that line will realize that one of the tunnel cuttings held an important clue to events around the Paleocene–Eocene boundary?

The fact that there were few if any thick *lignite* or coal seams found in the British Tertiary was a disappointment. There was a small deposit from the south of England that was known as the 'Cobham lignite'. This was a very thin and impersistent lignite that was found just to the south of London and was not well exposed in outcrop. So when reports were received that construction workers had found a thick lignite seam when cutting through sediments around the horizon of the lignite at Cobham, Margaret Collinson and colleagues rushed to the site. Working on temporary geological sites is not easy. In the case of archaeological sites, there is UK government legislation allowing a rescue dig that can take days, and sometimes even weeks or months. But geological sites do not have that protection. Indeed, it was fortunate that the contractors allowed any access at all. The geologists were given one day to record the site and take samples.

This was no mean task. They quickly realized the potential of the site. If the carbon isotope excursion was to be found anywhere in the British Isles in terrestrial rocks, it was here. But how to sample was an issue. We know that the isotope excursion represented a very short interval in time, and peat takes a considerable time to form. The excursion is therefore likely to be represented by a very thin horizon. Just grabbing a few bags of lignite was not going to be sufficient, but what to do? After their allocated day, the site was to be covered over in concrete! They were forced, therefore, to try to remove plaster-jacketed blocks of lignite from the side of the cutting and take them back to the laboratory for further study.

The priority was to describe the sequence of rocks and to undertake the isotopic analyses to see whether there was any indication of the carbon isotope excursion, and if so where it might lie. The excitement was also raised a level when some loose blocks of the lignite were collected and split, revealing fossil charcoal. At this point, with my experience of coal and charcoal, Margaret Collinson drew me into the project.

When the results of the isotope analyses came back, they indicated considerable perturbations in the carbon isotope record in the lower part of the lignite, which was in the form of fine layers or *laminations*, and contained distinctive charcoal layers. The uppermost part of the lignite was blocky and did not contain any recognizable charcoal. Had the carbon isotope excursion at the PETM been found in southern England?[5] And if so, what was the relationship of fire to how the peat was formed and to this period of rapid global warming?

With further isotopic work, we were able to confirm that there was a distinctive excursion within the laminated lignite. This was identified as the onset of the PETM. But to pick up clues as to what was going on during the period, we needed to examine the lignite under the microscope (*petrography*), and look at the distribution of the charcoal in more detail. In this, the quick thinking of the geologists in getting hold of large blocks proved helpful.

There was no point following the traditional method of coal petrographic analysis, in which coals are crushed and data averages obtained. After all, the isotope excursion was represented in the rock section by a very thin lignite unit. Instead, we made a continuous series of large polished blocks through the sequence, so that the exact position of any changes could be plotted. The decision to look at polished blocks of coal rather than crushed pellets was the right one, and not just because of the thinness of the layer associated with the carbon excursion. Some of the charcoal that had been recovered from the maceration of the coal looked as if it had come from ferns. But charcoal layers in coal can easily be crushed with the compaction of the peat when it is being transformed from peat to lignite, as the charcoal is much more brittle than the surrounding peaty matrix. Looking at the charcoal without extracting it from the lignite proved essential.

In those days, before photo-montage software was available, in order to image whole plant organs preserved as charcoal we had to take a large number of photographs with a low-magnification (×10) lens and then laboriously stitch them together using a computer drawing program. One of the images of a whole cross section of a fern leaf, or petiole, was the result of stitching together 56 individual photos—a real labour of love for our postdoctoral researcher, David Steart (BW Plate 11).[6] Not surprisingly, up until then few coal petrologists had appreciated that large plant organs may be preserved in this way.

Our petrographic study confirmed that charcoal was only common in the laminated lignite part of the coal that occurred in the lower part of the section, where the isotope excursion had been found. Ferns dominated the charcoal, although some angiosperm wood also occurred. The charcoal appeared in bands, suggesting a large number of fire events that were followed by post-fire erosion.[7] Ferns are typical recolonizing plants, and fire may have been frequent enough to prevent the growth and domination of trees in the ecosystem. But we found no evidence of burned peat, which we may have expected if this was a cause of the carbon excursion.

In contrast, there was little or no charcoal found in the blocky lignite above the onset of the PETM. This coal was formed as the result of the decay of herbaceous plants, as it was rich in decayed leaves, cuticles, and herbaceous tissue. The transition was interpreted as indicating a change in hydrology (water regime), with the loss of fire, increased run-off, and waterlogging resulting from increased rainfall linked to the onset of the PETM.[8] But was there any other evidence of increased rainfall at this time, or was this simply a local change? David Beerling, from the University of Sheffield, who is always keen to model alternative scenarios, had shown in one of his model runs that we might expect increased rainfall at the onset and during the PETM, so perhaps here was some supporting evidence.[9]

We could find out about any related change in vegetation in our British rock sequences by examining the microscopic spore and pollen record. This confirmed what we had learned from the laminated lignite—fern spores dominated, as would have been expected.[10] These were very distinctive spores called *Ciccatricosisisporites* and were probably produced by Schizaceous ferns. The vegetation of the PETM interval itself was characterized by the loss of ferns and cessation of fires, and an increase in wetland plants including conifers of the Cypress family (Cupressaceae) and a more varied flowering plant community that also included palms.[11] Unfortunately we do not know what types of palms these were, but I like to think of southern England as once a warm tropical paradise with palms swaying in the wind! The upper blocky lignite was calculated to represent 4,000–12,000 years after the onset of the PETM, and the flora becomes richer but in many ways is very similar to the flora recorded below. At Cobham we cannot see what happened next, as the succeeding environments are very different and do not preserve the spores and pollen very well. They also include marine sediments.

If there were changes in the environment, fire systems, vegetation, and water conditions during the PETM, what could we learn by looking at the organic compounds extracted from the coal? Could they hold clues to the cause or effects of the PETM? We enlisted the help of organic geochemists at the University of Bristol, led by Rich Pancost, who looked for biomarkers. These are chemicals that have been derived from known organisms, and so indicate their former presence in the rock. One such group of compounds, the hopanoids, are derived from bacteria. The Bristol group was able to trace not only the distribution of hopanoids in our samples but also the changes to their carbon isotopic values, which would tell us much about the environmental events that were occurring during this important interval. We were particularly interested to learn about the changes in the wetland ecosystems at the time.

As I mentioned earlier, there had been a suggestion of a rapid rise in the potent greenhouse gas methane in the atmosphere, causing some of the global warming. Although some of this may have been from the release of marine methane hydrates, there was also the possibility that some methane may have been released from high-latitude wetlands, but direct evidence was lacking. The carbon isotopic values of the hopanoids in the Cobham rocks at the onset of the PETM indicated that there had been an increase in the population of bacteria which feed on methane (methanotrophs) at this time. This was evidence for an increase in methane production, potentially driven by changes to a warmer and wetter climate, as climate models had suggested. Methane release from the land could have acted as a positive feedback mechanism, by which increasingly wetter climates led to more methane release, which in turn increased global warming.[12]

By 55 million years ago, the atmospheric oxygen concentration appears to have stabilized at the modern level of 21 per cent. Fires would have been controlled more by rainfall than oxygen level, as they are today, and wetter climates would have led to a decrease in fire activity. There had already been some research that indicated that the Eocene climate became wetter and more monsoonal, and that tropical rainforests had evolved and were spreading widely. As we know from the present day, fires are not a natural part of the tropical rainforest system. Where could we find additional data?

A good locality to obtain further evidence was an open pit coal mine at Schöningen, which, curiously, had once straddled the border between West and East Germany. Even today, some of the border fence is still preserved next to the mine. The coal sequence was known to include the Paleocene–Eocene boundary and extend into the early Eocene, and charcoal was known to be more abundant in some coals than others.[13] Here we could build up the picture up to the end of the early Eocene (about 48 million years ago), well beyond where our Cobham lignite left off.

Standing on the edge of the open pit mine, the scale was difficult to comprehend (Figure 48). The coalmine exposes 11 coal seams, with the lower, main seam being around 10 metres thick. The overall coal-bearing sequence is around 170 metres thick. This coal sequence was deposited on the margins of a tropical sea in the northern hemisphere. Indeed, we saw rooted palm

Figure 48. The Eocene (55–50 myr) lignites (brown coal—dark) from Schöningen, Germany. The mine exposes several thick lignite layers.

trees in the top of the sequence. The rocks represented three main sequences that were cyclically flooded by the sea, with the low coastal areas returning to peat formation after flooding.

It was a long walk down to the main seam to get rock samples—three lots of them, for petrography, for the isotope geochemistry, and for palynology—to identify vegetation from microscopic pollen and spores. It was hard and often dirty work, but we collected a significant amount of material. Fortunately we had arranged for at least some of it to be driven out of the mine.

As I write this, our research is not yet complete, but we have made progress. New ways to interpret the temperature at the time have indicated that lands at 48°N in this part of the early Eocene experienced mean annual temperatures ranging from 23°C to 26°C.[14] The evidence from Schöningen indicates that fire was significant in the lower part of the sequence. In general, although charcoal abundance was low relative to previous high-fire worlds such as the Cretaceous, wildfire activity was higher than today during the warmest parts of the Paleogene—that is, during the PETM interval covering the end of the Paleocene and earliest part of the Eocene.[15] By this time, as we noted, atmospheric oxygen levels had stabilized to modern values, and precipitation and humidity became the main control on wildfire. Increased rainfall, encouraging lush vegetation, followed by droughts, would have created an environment rich in dry fuel in which wildfires could easily propagate if humidity became low enough.

A few million years into the Eocene, by around 50 million years ago, charcoal abundance falls to levels similar to those found in modern peats. The transition to the modern low-fire world appears to have occurred within the early Eocene, and records from across the globe confirm that by 45 million years ago wildfire systems operated much as they do today (Figure 49).

For the time following the Eocene, there is little information available, but we are beginning to see isolated records, and charcoal contents of coals are seen recorded as inertinite, indicating that there was a normal fire world—not one without fire. Our lack of evidence of fire for this time period is, therefore, due to a lack of searching rather than a true absence. I remember more than 40 years ago first visiting the large open pit coal mines near Cologne that mine lignites more than 50 metres thick, formed in the Miocene Period around 15 million years ago, and collecting charcoal from within the coal seams. Granted, the charcoal was not as abundant as in the Carboniferous coals, but it was there and certainly obvious. Most recently, evidence of fires has come from the Oligocene–Miocene (34–8 million years ago) coals of Australia, and the researchers suggest that some of the swamp plants were adapted to fire and may have been the ancestors of the fire-adapted flora that we see in Australia today.[16]

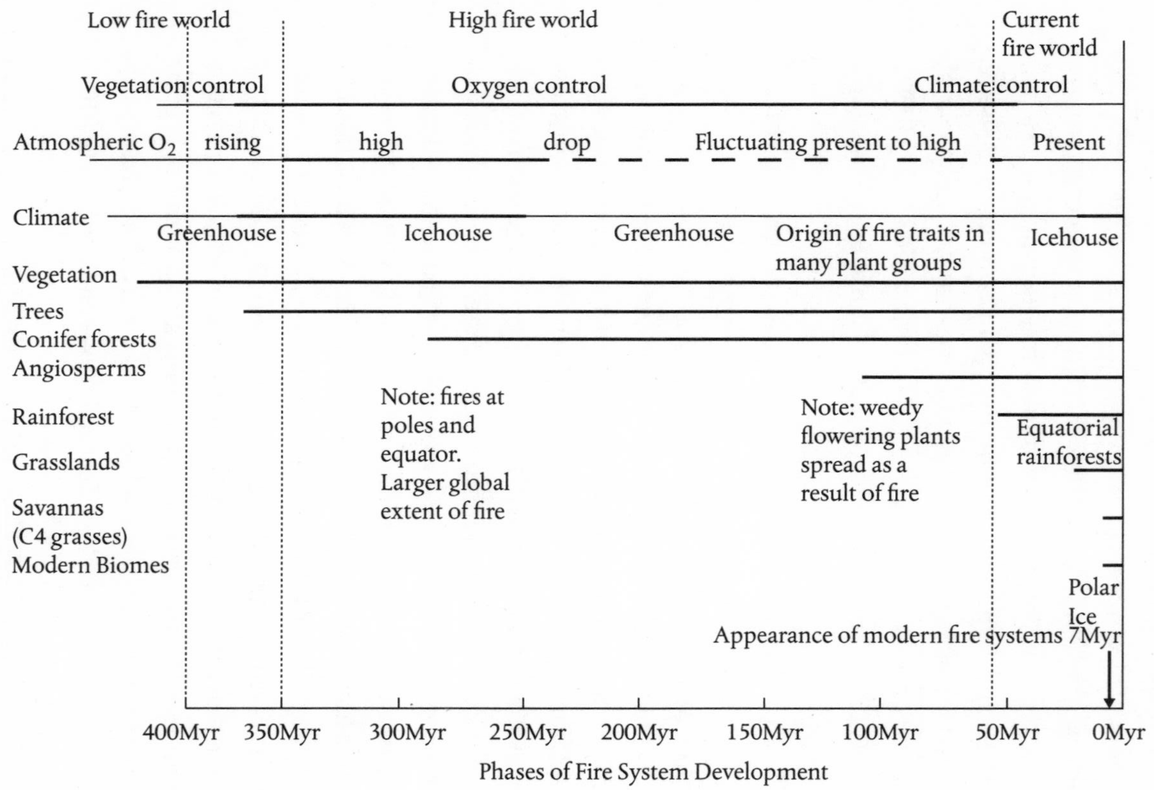

Figure 49. Phases of fire system development through geological time, showing the major innovations in plant evolution and showing the change from a low-fire to high-fire world, and to the current fire world.

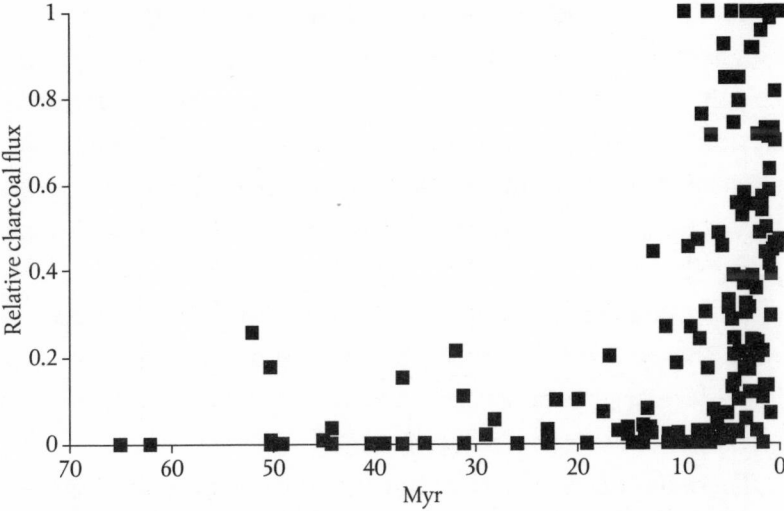

Figure 50. Increasing charcoal flux in oceanic sediments, reflecting the rise of fire in savanna grasslands.

If we are to get a more continuous picture perhaps we should look to marine cores. As we have seen, charcoal may be carried some distance in the wind, and this may provide us with at least some record of biomass burning. The only major study of this type was by an American researcher, J.R. Herring. In 1985 he published an important work looking at a series of cores in the Pacific and Atlantic oceans. He showed that there was a significant record of fire in these marine sediments covering the past 50 million years.[17] He was also able to show that the nature of the record changed, with an increase in fire over the past 7 million years or so (Figure 50). Subsequent work also indicates a major change in fire activity over the past 10 million years. Why was this so?

The impact of grasslands

It seems that the world's climate generally became drier over the past 30 million years or so. One of our first clues as to what was happening to vegetation during this period came from the carbon isotope record. A landmark paper by Thur Cerling from the USA in 1993[18] showed that there was a major shift in terrestrial carbon isotopic values around 7 million years ago. We know that plants with different metabolic pathways select between the carbon isotopes (fractionate) in a different way, so that the traditional plants that use a C3 metabolic pathway have a different value of the isotope ratio, $^{13}C/^{12}C$, to those plants that evolved a C4 pathway. The evolution of the C4 pathway allowed angiosperms to flourish in drier habitats, and is a feature of many of the grasses seen in drier environments today.

There has been a flurry of interest in the evolution of grasses and grasslands, and it is hard to believe that they appeared relatively recently in Earth history. Grasslands (as opposed to individual grass species) probably originated in the Oligocene Epoch, not much more than 30 million years ago. The C4 grasslands spread during the subsequent Miocene and Pliocene to form large areas of savanna. And one of the possible causes for the spread and persistence of the savanna biome is fire.[19]

One of the most visually compelling arguments for fire as a causative factor comes from the modelling of vegetation in a world without fire.[20] Such modelling indicates that if fire were excluded then the savannas would disappear and revert to scrub or forest. Further, if there were regular fires, then the flame height in grasslands would be sufficient to kill the seedlings of trees and shrubs. But if the interval between fire events was greater than a decade, then the growing tips of the seedlings could rise above the flames, allowing them to survive and leading to replacement of the grassland by scrubland, and eventually forest. The rise of savannas played a part in our human evolution story as well.

From 10 million years ago, the global temperature dropped, taking the Earth out of the greenhouse world that it had experienced for more than 200 million years to an 'icehouse' world. There is evidence of vegetation in Antarctica even up to 2 million years ago.[21] The expansion of the Antarctic ice sheet is, therefore, relatively recent and may have been partly the result of the opening of the Drake Passage and the change in the circulation of the oceans, as well as the growth of the Himalaya by the continued collision of India into Asia. [22]

Decoding the recent history of fire

Much of our search for fire in the fossil record has been tied up with human evolution, as it is this species alone that has learned to control fire. We will examine this in Chapter 7, but here I want to consider how we can unravel the more recent part of our fire history. Until now, our story has been based on two main forms of evidence: macroscopic charcoal that can be found in sediments, and charcoal that is found in coal that has been recorded as inertinite macerals, including fusinite and semifusinite. As we come up the geological column towards the present day, the amount of data increases considerably. The main concern in this most recent interval is to understand the impact of the plunge into the ice ages. For this we need to integrate studies of climate and vegetational change.

The most obvious way is to look at peat and lake cores and examine the spores and pollen of the plants that were living in the area. As we reach more recent times, many of the plants familiar to us appear, and the distribution and quantity of these plants over the past million years or so appears to have been controlled by climate. In order to understand the synchroneity of past events, an accurate age model is required for the sequences being studied. In deep time, it is sufficient to consider changes on the million-year or often the 10-million-year timescale, but when we come to studying the recent past, we need to think in terms of thousands of years, not millions.

Charcoal particles can be counted on thin sections made from polished rock, but interpretation is not straightforward. How can one level or horizon be compared with another, as we are unsure what time the unit represents? And does the increase in charcoal observed in the past 7 million years indicate more fire, or just a change in the way in which the charcoal accumulates and is preserved?

There are a number of ways in which the charcoal is obtained, studied, and quantified, with which some take issue, but overall the approach has revealed plausible results. One of the problems is, what constitutes a charcoal particle? We see from modern fires that they can produce charcoal at all scale ranges, from microns (10^{-6} m) to over 1 cm in any dimension. When a large charcoal fragment is buried, especially if it was produced at high temperature, it will break up more readily as the sediment compacts with time, as well as potentially during the processing of the samples. We could consider charcoal volume, but again from our modern fire studies we find the volume of charcoal produced can be quite variable. There is also the assumption that large charcoal fragments are more indicative of a nearby fire. If we were dealing with only wind-blown charcoal this may indeed be the case, but as we have seen, in water larger pieces of charcoal can take longer to sink and can travel much further.

In spite of these problems, charcoal data have been recorded predominantly from sequences that were formed in the same environment and by the same process. If we have, for example, a lake that was accumulating sediment at a steady rate, or at least rates that can be accurately determined, the sudden increase in charcoal over the background can be used to interpret a fire event, at least locally. We can apply the same approach to peat deposits.

Another difficulty is how to compare data from different parts of the world, which have been studied by researchers who have used slightly different methods to obtain those data. A working group of researchers was set up to resolve these issues, and this resulted in a global charcoal database mainly covering the past 70,000 years. Data are being added on a daily basis and the database is now available online for all to use.[23] The data are transformed mathematically to allow records to be compared, and when considered together these transformed data can be used to address changes in fire history on a regional or even global basis (Figure 51). Unfortunately this information does not help solve all the issues concerning the history of fire over that period. Really, we would also like to know what vegetation is burning at any one place and the links between climate change, vegetation change, and indeed human-induced change through land clearance and

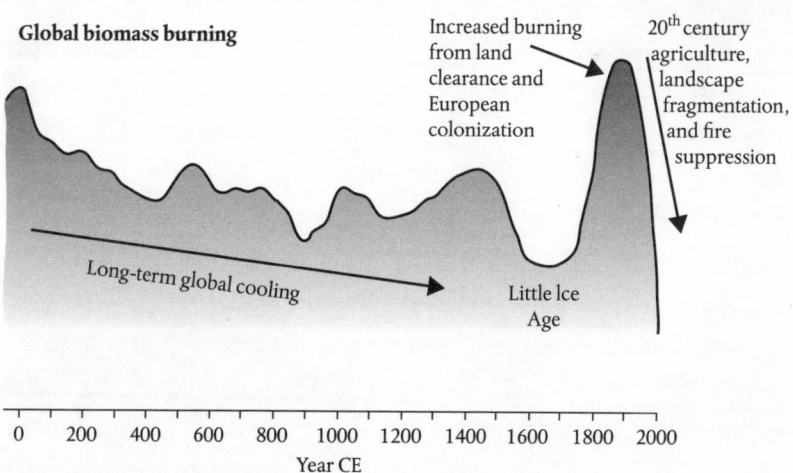

Figure 51. Wildfire over the past 2,000 years, showing a marked change in the twentieth century.

variations in land use. Perhaps the advances we have made in the study of macroscopic charcoal, which allow us to discover what has been burned, will help in the future.

There are other clues that can reveal recent fire history. When a fire passes through a forest, as we noted, many of the trees will not be killed. Often the fire burns through just one side of the base of the trunk and growth later returns. When the tree is cut down, in addition to the tree rings we can see scars left by the fire. Because it has been possible to establish a tree ring chronology, we can use it to help interpret fire history on a year-on-year basis, not only within a single forest, but across much wider regions (Figure 52). The beauty of this technique is that fire events can be linked to changes in climate that are also recorded within the tree rings as changes in ring width, and also in the isotopic composition of the carbon that makes up the wood.

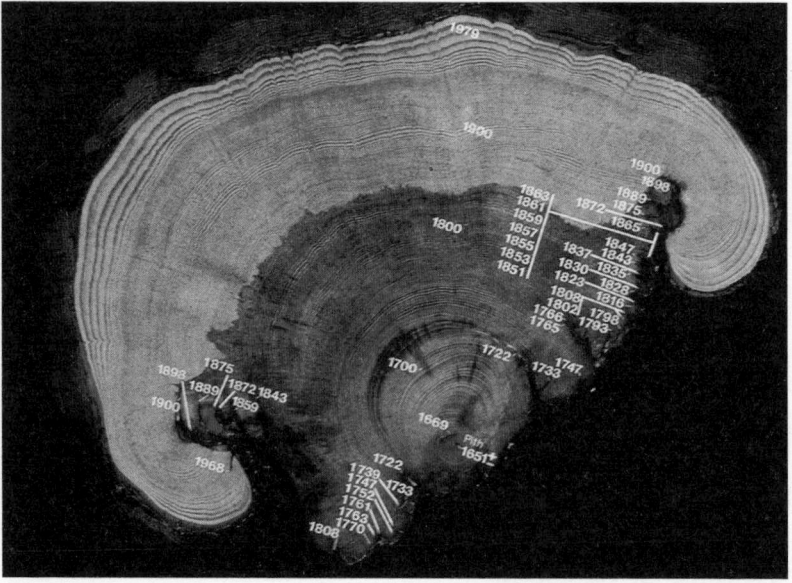

Figure 52. Cross section of coniferous tree trunk showing 42 fire scars over 400 years.

Some of the oldest fire scar/tree ring histories have been obtained from the Sequoia National Park in western North America.[24] It has been possible to develop a 3,000-year history of fire based upon these data. Fire data have also been obtained from lake sequences using measurements of micro-charcoal, and in this area it has been possible to integrate the two datasets. These studies show that a regular surface fire regime persisted for many millennia. In recent years, by contrast, fire suppression has increased in many areas in the western USA, so that when fires do occur, they are much more severe and turn from surface fires into much more severe crown fires, resulting in the loss of many more trees and a greater impact on the community structure.

The development of fire scar/tree ring chronologies in the western USA has proved important in examining not only changes in natural versus human activity in the start and spread of fire, but also in the study of fire and climate change. One of the early key results was that as temperatures rise there is an increase in fire activity.[25] And fire activity is also affected by the large-scale phenomena known as El Niño and El Niña. The development of the charcoal database, too, has provided insights into the relationship between climate change and fire. One interesting result is the discovery that there is often increased incidence of fire during periods of rapid climate change.[26] We will return to the implications for our present situation at the end of the book.

The Younger Dryas impact hypothesis

Just as the idea of continent-wide fires had been questioned with studies on the K/P boundary, in 2009 we were engulfed again with another claim of impacts and fires. This time the claim, and my own involvement in the debate, began in 2006–7, while I was on sabbatical at Yale University. I had become involved in a study of fire history in the Northern Channel Islands off the coast of Santa Barbara in California. This is where some of the oldest human remains in North America have been found, so it was a good locality to investigate fire history before human arrival, and then see the impact of human activity on the vegetation–fire system.

I was aware that charcoal was present in the sediments, as carbon dating had been used on the remains, and I was initially sent material from a site named AC003, upstream from the human site in Arlington Canyon on Santa Rosa Island in early 2007. It was at this time that reports appeared of an announcement at a scientific meeting, by a large group of researchers, that they had evidence of a major comet impact in North America that occurred about 13,000 years ago. They claimed that the impact caused a continent-wide fire which resulted in the extinction of the early human Clovis culture in the continent, as well as the extinction of all the larger animals in North America, including the mammoths, and a major global climate shift, with the onset of the cold period known as the Younger Dryas.[27] The media quickly took up the idea, and several special TV documentaries were produced. My heart sank. Science requires time for examination and assessment of the evidence by a number of groups, but in the meantime, a dramatic scenario which captures the imagination of the media can become popular mythology and hard to shift, as has happened with the idea of a global fire at the K/P extinction event.

The research was published later, in a prestigious journal.[28] With such a lot of data and so many authors, some of whom were famous, how could this hypothesis not be true? But I was already puzzled by the claim for a continent-wide fire, based on data I regarded as problematic. Subsequently the same authors used data from the charcoal, and what they called carbonaceous spherules and elongates, from the site AC003 in Arlington, to argue further for these impact-started high-intensity fires.[29] I had already begun work on the charcoal in the AC003 samples sent to me in 2007, and it was clear to me that this was a normal charcoal assemblage. It contained abundant wood charcoal and a range of other objects, including arthropod faecal pellets. There were also carbonaceous spherules, whose origin I was unsure of at the time, but I had often seen similar spherules in modern fire residues.

It was time to examine the field evidence for myself. In 2008 I went with American colleagues to Santa Cruz Island, another of the California Channel Islands, to start our research. We had also planned to work on Santa Rosa, where the AC003 samples had originated, but bad weather prevented us, though we managed to reach there in subsequent years. We found many sites across the islands with charcoal, and a range of the other objects that had attracted so much attention (Figure 53).

We were baffled by the analysis and interpretation of the charcoal by the impact group. Rather than high-temperature wildfires, our reflectance studies indicated that the fires on Santa Rosa had been mainly low-temperature surface fires.[30]

The question of the identity of the carbon elongates and carbonaceous spherules became increasingly important in the debate. Some of these were claimed to contain nanodiamonds, which would imply an impact.[31] However, from my own knowledge of studying arthropod faecal pellets in the fossil record, the elongates were dead ringers for arthropod faecal material, and some were even identical to frass from termites.[32]

The carbonaceous spherules were another problem. I had found these in my modern fire samples and from the fire at Frensham, and I knew, as I was there, that it was a low-temperature surface

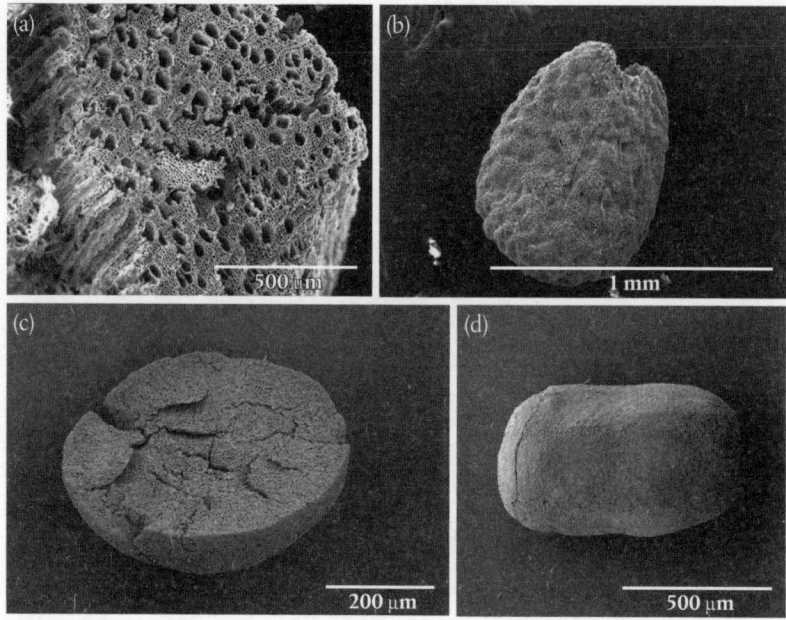

Figure 53. Scanning electron micrographs of charcoal in late Pleistocene sediments around 13,000 years ago, in fluvial sandstones, Arlington Canyon, Santa Rosa, California Channel Islands, USA. (a) Angiosperm wood with vessels, and (b) seed. (c) Fungal sclerotia and (d) termite coprolites (faecal pellets) have caused considerable confusion and have been erroneously claimed to represent carbonaceous spherules and elongates related to high-temperature fires following an impact.

fire. There had been no comet impact on southern England! From discussions with colleagues we suspected that these spherules were in fact fungal sclerotia—round, hardened masses of fungal mycelium. These are formed when soil fungi are under stress—they are a kind of resting stage for the fungus (not spores as some have mistakenly claimed). They are found globally in soils, and particularly in soils that have regularly been subjected to fire. I needed to look at some material. Down the road from Royal Holloway was the Institute of Mycology, and I made contact with researchers there to look at fungal sclerotia, including examples charred at different temperatures. It became clear to me that a large number of the carbonaceous spherules or spheres were indeed fungal sclerotia, and had not been formed by high-temperature, continent-wide crown fires.[33] We also cast doubt on the nanodiamond story.[34]

The impact hypothesis remains under debate. From my point of view, I do not think that there was an impact. But no doubt it will resurface in the media for many years to come!

Prometheus

We are uniquely fire creatures on a uniquely fire planet.

S.J. Pyne

It is sometimes said that humans were born of fire. While a wide range of animal species interact with fire, we appear to be the only species to have learned to tame it, and more importantly to make it (Figure 54). There is evidence that early humans were aware of fire and may have exploited naturally occurring fire, but only later did they control and manage it.

Human interaction with fire must have proceeded through various levels, the first of which can be described as the opportunistic phase (Figure 55). In this phase, natural fire may have been exploited to help in hunting, for example. When, how, and why did this happen? It is widely agreed that our story begins in Africa. It is here that we see the evolution of hominins, a group of related genera that include the Australopithecines and later the genus *Homo*. How common would fire have been in the environments in which they lived?

We already know from the study of fossil plants, as well as isotope data, that there were important changes in both the vegetation and climate over the past 10 million years. It is also during this time interval that hominins emerged from apes. Through the Oligocene and Miocene (30–8 million years ago), Africa was largely covered by tropical rainforest, where fire was present but

Figure 54. Early Fire Experiments by Randall McIlwaine.

infrequent, started both by lightning strikes and volcanic activity. As the climate began to dry and C4 grasses spread at the end of the Miocene Epoch, around 8 million years ago, habitats became more open. Fire became more frequent, and from an animal perspective would have become more visible, not just from flames but also smoke. Frequent fire in the landscape would have had many consequences for the early hominins, not just because game was more easily killed, but burned animals (naturally cooked meat) would have made a useful addition to the diet, and the new flush of growth following fire would also have attracted

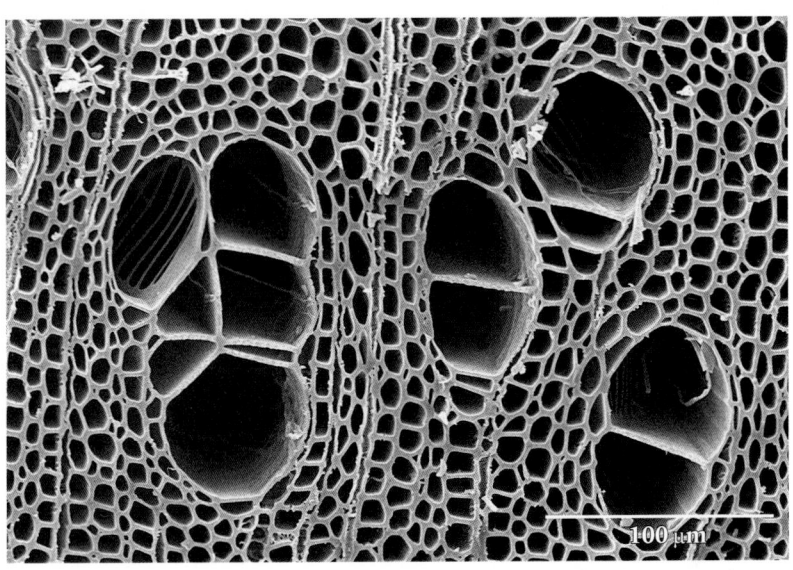

Plate 1. Scanning electron micrograph of *Betula* charcoal showing good anatomical preservation. The smaller cells are tracheids (water-transporting cells of the xylem), and the larger cells are vessels.

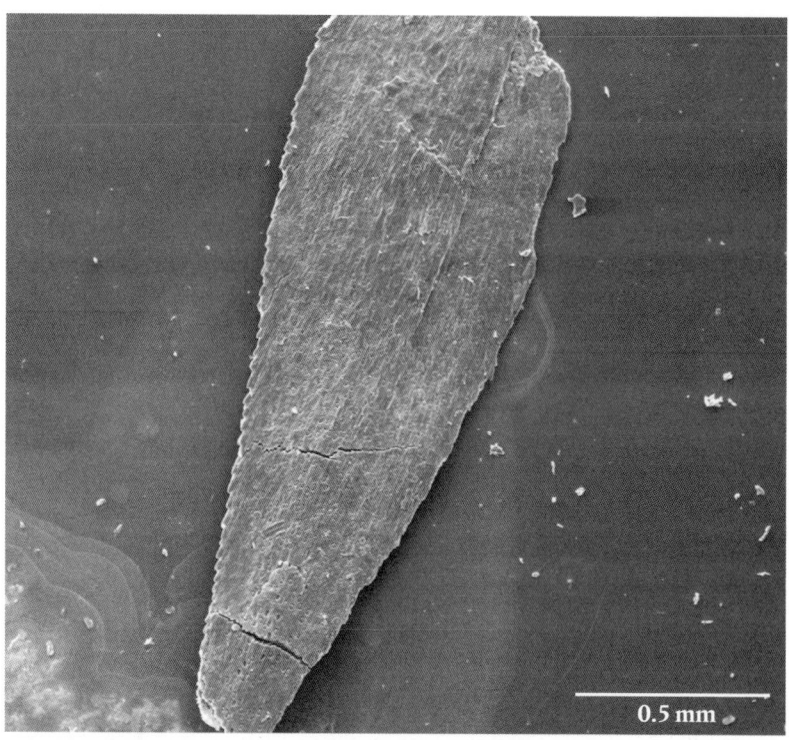

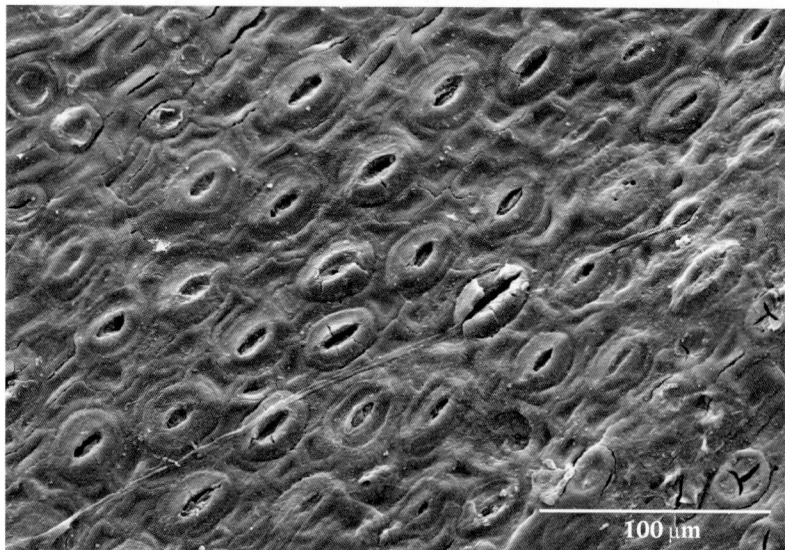

Plate 2. Scanning electron micrograph of the earliest fossil conifer, Carboniferous (300 myr), *Swillingtonia* from Yorkshire, England. (a) whole leaf; (b) detail of stomata.

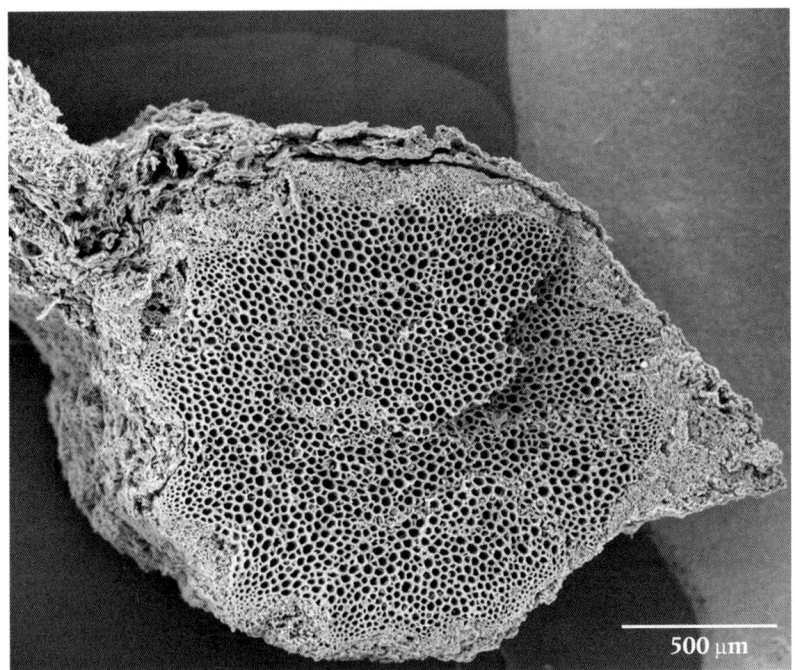

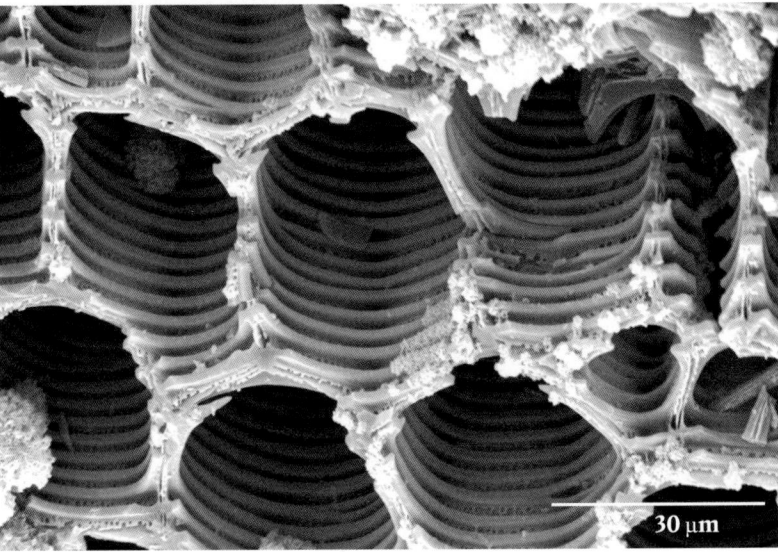

Plate 3. Scanning electron micrographs of the charcoalified stele (the central part of the stem) of the early herbaceous lycopsid plant *Oxroadia* from the early Carboniferous sequence (325 myr) at Shalwy, Donegal, Ireland. The specimen was dissolved from the rock using acid. (a) Whole stele; (b) detail of tracheids showing thickening bars.

Plate 4. (a) Scanning electron micrograph of 335 million-year-old (Carboniferous) charcoalified pollen organ from Scotland; (b) Synchrotron X-ray Microscopic Tomographic image of the same specimen showing a virtual section; and (c) a reconstructed image.

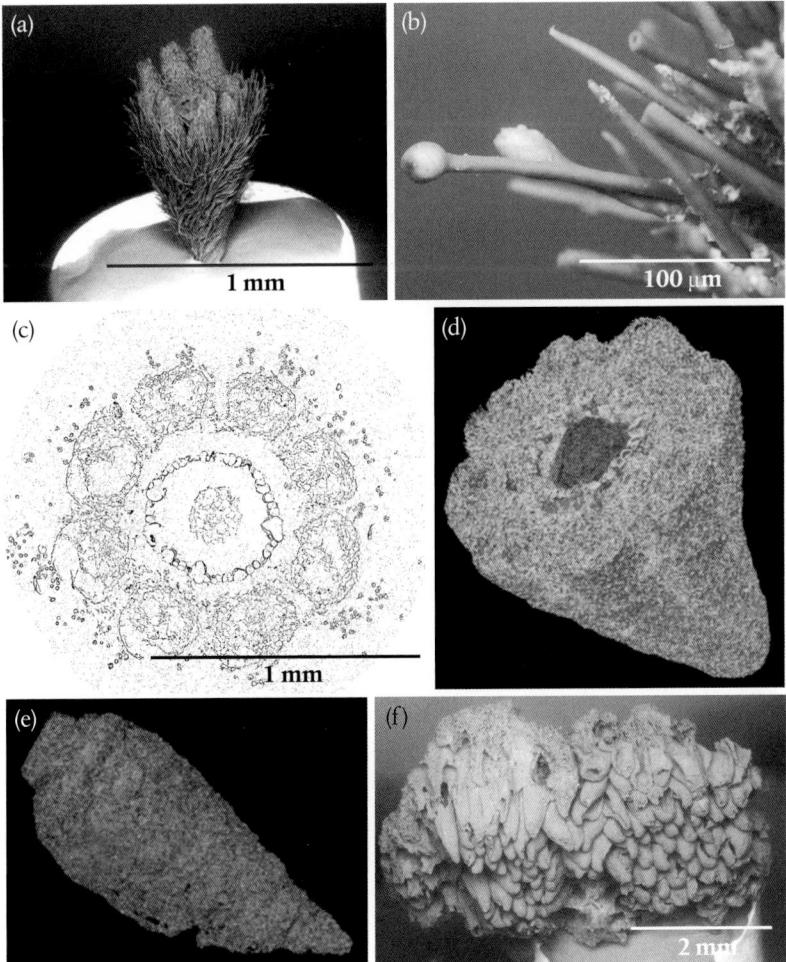

Plate 5. Exceptionally preserved charcoalified fertile organs of 335 million-year-old seed-ferns from the Kingswood limestone, Scotland, can be studied using many techniques. Using the SEM, details of (a) a 1 mm-long ovule can be seen, with (b) spirally arranged glandular hairs. Using synchrotron radiation X-ray tomography, (c) the internal cross-section can be digitally imaged non-destructively. Many such slices can be used to reconstruct (d) the ovule with colour coding of different layers that can be digitally stripped away, showing for example (e) the internal megaspore. Such techniques have also been used to study (f) pollen organs from the same deposit.

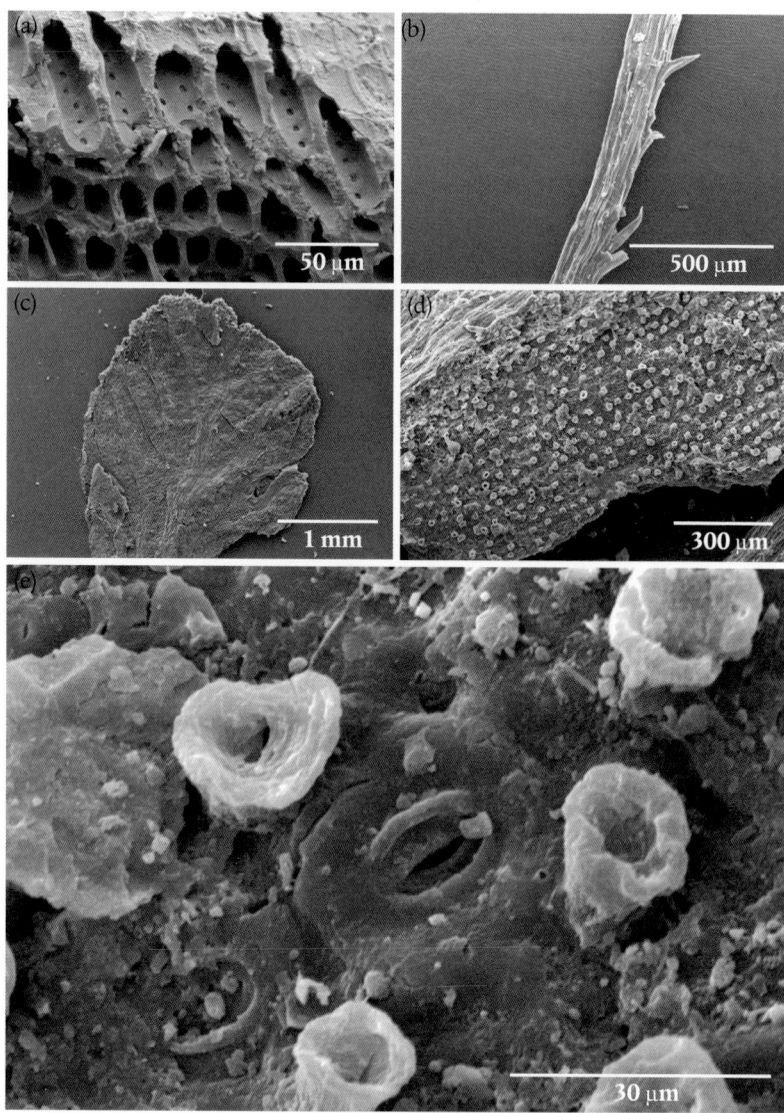

Plate 6. Plant fragments including charcoal from Carboniferous (310 myr) coal measure sediments, Swillington Quarry, near Leeds, Yorkshire, UK. Scanning electron micrographs of a range of plants and plant parts showing beautiful anatomical preservation: (a) cordaite wood; (b) spiny axis; (c) Pteridosperm leaf; (d) cordaite leaf; (e) detail of (d) showing stoma and papillae.

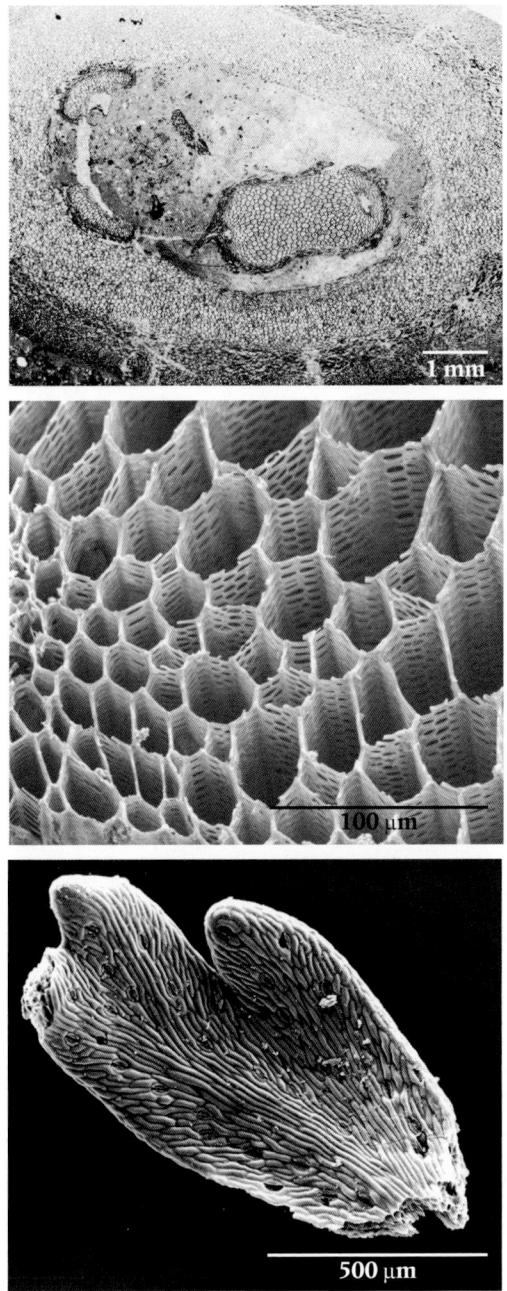

Plate 7. (a) Black charcoal fragments in limestone from the Lower Carboniferous (335 myr) of Pettycur, Scotland; (b) scanning electron micrograph of the main stem of a fern (rachis); and (c) a seed-fern leaf with well preserved stomata. The density of stomata can yield data on the carbon dioxide content of the atmosphere.

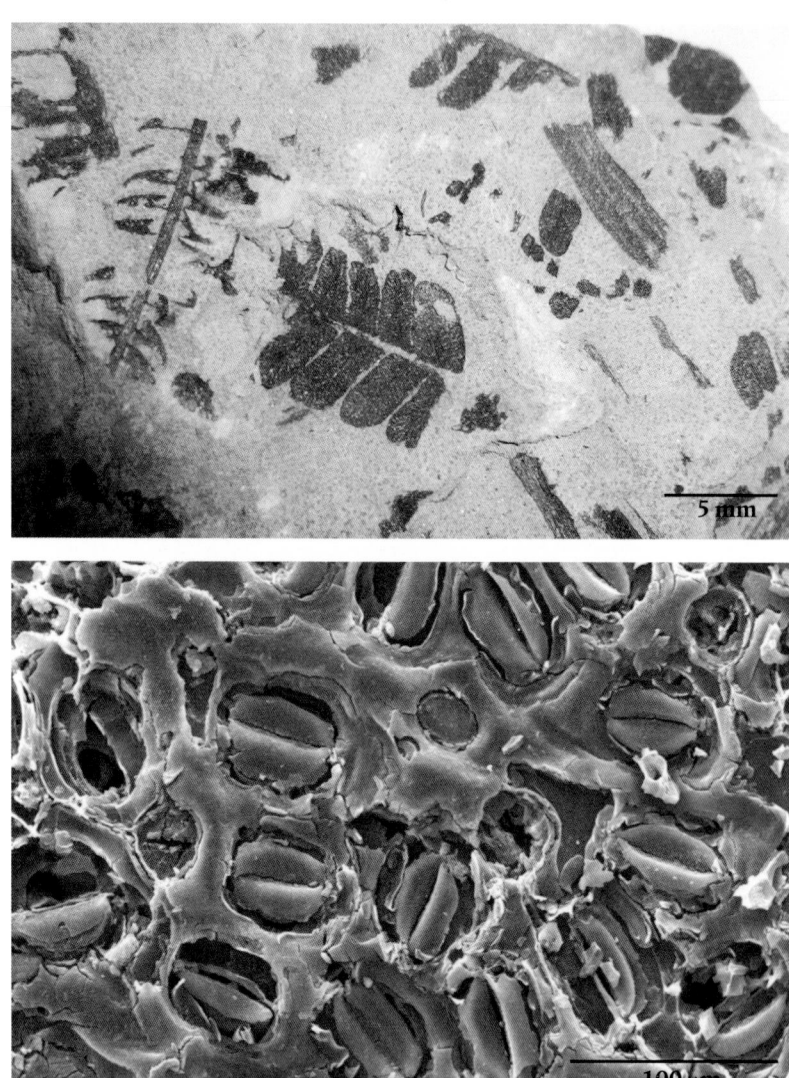

Plate 8. (a) Charcoal from the Lower Cretaceous (120 myr) Wealden of the Isle of Wight, England, including charcoalified ferns; (b) their anatomy can be studied using scanning electron microscopy. The undersurface of the leaf in (b) shows numerous stomata (gas exchange pores).

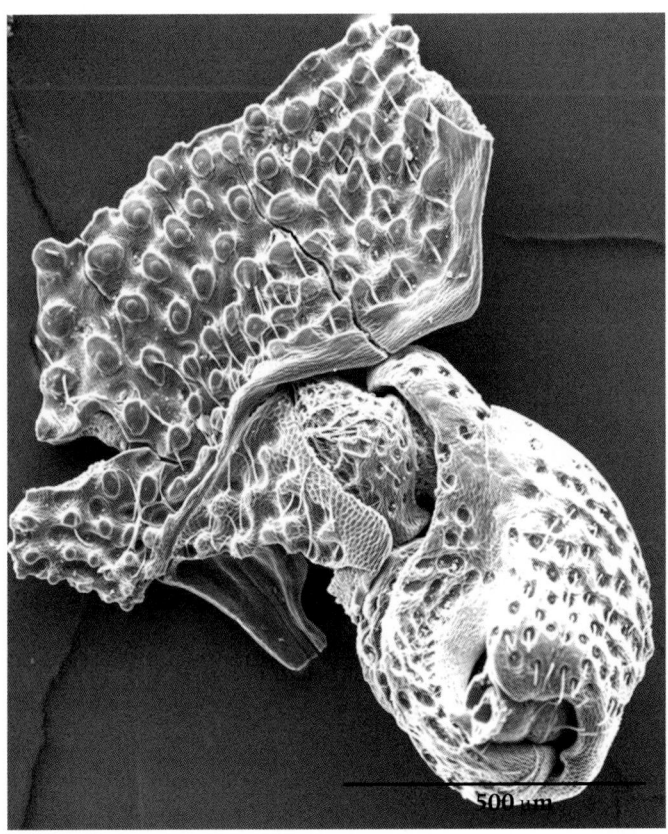

Plate 9. Charcoal from the Wealden sediments, Lower Cretaceous (120 myr) of Belgium. Thick charred litter layers can be found that yield small charcoalified plants but also charred insect fragments seen using the SEM, such as the one illustrated here.

Plate 10. (a–c) Scanning electron micrograph of Cretaceous charcoalified flowers (110 myr), Georgia, USA; and (d) *Scandianthus* from Sweden (90 myr).

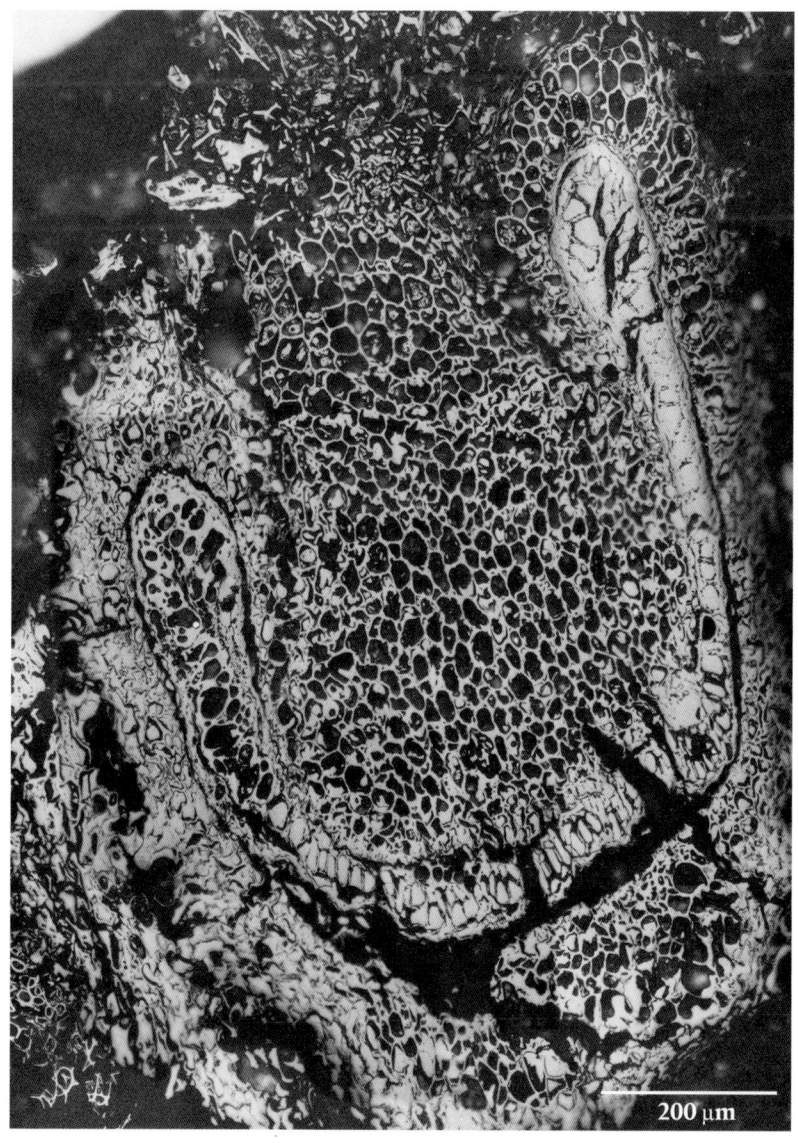

200 μm

Plate 11. Photo montage of 56 reflected light images under oil of coal block showing charcoalified fern rachis from the Cobham lignite, (55 myr), Kent, England.

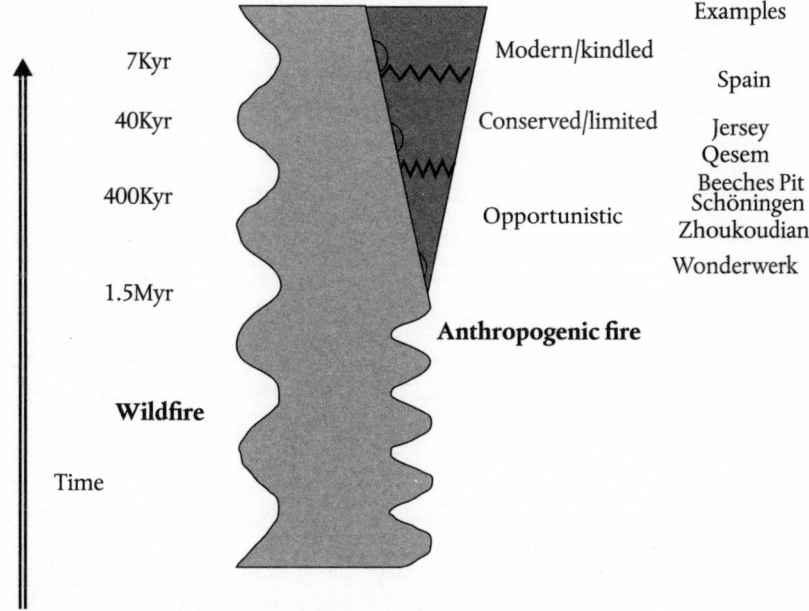

Figure 55. Wildfire and anthropogenic fire. There is a change from opportunistic to intentional use of fire.

large herds of herbivores. Fire may have been conserved through adding fuel, including dung, which is slow burning. An additional benefit of maintaining fire is for protection from animal attacks at night, and the smoke can also ward off insects. The ability to 'stretch' fire both in time and space appears to be a trait only exhibited by hominins.

Our exploitation of fire has been highly significant. It enabled us to expand our diet through cooking, which in turn is thought to have had an important impact on brain size, which increased through the Pleistocene.[1] The role of fire as a social focus may also have helped the development of language. John Gowlett has discussed both the cooking hypothesis and the social brain hypothesis with regard to the course of human evolution. His listing of major benefits of controlled fire to early humans is given in Table 1.[2]

Table 1. The major benefits of fire.

Protection	Against large predators
Warmth	Especially in high latitudes
Food preparation	Especially cooking meat and starch
Tool preparation	Especially of stone or wood, but also mastic and all later pyrotechnologies
Social focus	Group interactions, ritual, language

Earliest evidence of fire use

It has been suggested that chimps have the mental capacities needed to cook food,[3] and that has led to the idea that humans may have developed the ability to cook very soon after they learned to control fire (Figure 56). Many researchers now consider that it was *Homo erectus*, which had evolved in Africa by about 1.9 million years ago, that was the first to control and use fire.[4] *H. erectus* spread widely across Eurasia from Africa, and the ability to start and control fire would have been crucial to populations migrating north, to cooler, temperate climates, especially as conditions changed through the glacial–interglacial cycles. Even in the warmer interglacials the temperatures at night may have fallen quite considerably. Evidence of fire use by *H. erectus* has also been claimed from China.[5] And controlled fire use by Neanderthals has also been established, dating back to about 400,000 years ago.[6]

To obtain definite evidence of fire use from the normal rock record is rather difficult, and we shall discuss the possibilities in what follows, but the clearest evidence would be to find fire where there is evidence of the presence of humans and where fire would not naturally occur. The most obvious place to look is in

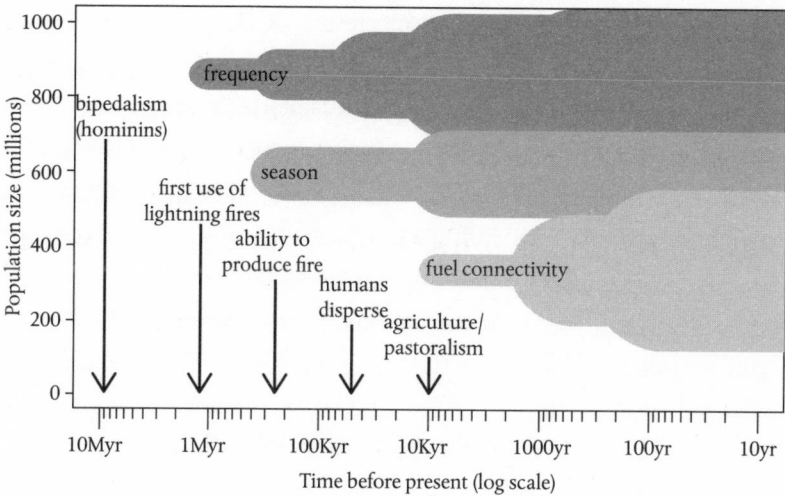

Figure 56. Fire and human activity. Stages of human evolution have been defined by the ability to manipulate the frequency and timing of ignition events.

cave sites. Claims have been varied—from evidence of charcoal occurrence to reddened flints and soils—but in many cases these are suggestive and not definitive.[7]

So what would constitute robust evidence of the deliberate use of fire? First, the site would have to be a cave that humans definitely inhabited. Looking at the context of any bones or tools is important, in order to remove any possibility that they may have been transported into the cave site by, say, water. Second, there should be evidence of fire confined to a specific place, and not in the form of scattered charcoal in the sediment. A well-defined hearth would be useful. Third, the hearth should reveal evidence of raised temperatures, not only through the presence of charcoal but baked sediment. If we are in luck, we may just find such evidence, but it is not going to be easy. And when searching for the first human use of fire, we need to look in the cradle of humankind—in Africa.

Archaeologists looking for evidence of the early use of fire often look for the presence of a hearth. But how can we identify a hearth in the fossil/archaeological record? A hearth has been defined as 'remnants of a domestic fire feature that retains some or most of its original structure or compositional elements (e.g. organic matter and overlying ash)',[8] but it is not straightforward to specify, and there have been many attempts to define and study hearths.[9] Much research has been conducted in Africa seeking the early use of fire, and here we will just look at some of the more recent work.

One piece of evidence comes from a cave in southern Africa. The Wonderwerk Cave is in the Northern Cape Province of South Africa and the sediments it contains date from around 1 million years ago.[10] The researchers were attempting to find definite evidence of fire being used in situ within the cave. It was clear that early hominins had used the cave: a number of flint axe heads were discovered at various positions. The evidence of fire comes from one particular level. The team used a new technique, called MFTIR (micro-morphological Fourier transform infra-red micro-spectroscopy), to provide unambiguous evidence of burning, both of the sediment and associated bones. This tech-nique uses infrared light to probe samples. The amount of infra-red absorption and transmission is related to the type of material and the temperature to which it has been subjected. In our search for evidence of human fire use, 1 million years ago seems at pre-sent a good marker point. While we are forever searching for the earliest occurrence, evidence in the deeper time record is scant. For increasing evidence of fire use, we have to look to an interval within the Lower Palaeolithic, between 400,000 and 300,000 years ago.[11]

Evidence of the habitual use of fire has been claimed in Europe from this time.[12] This evidence, in the form of heated flints and the charred tip from a spear, comes from Schöningen in Germany (where we had described the evidence for wildfire from 40 million years ago). Evidence for habitual fire use has also been claimed from the Levant, dated between 350,000 and 320,000 years ago.[13] Some of these data come from the Tabun VI and Qesem caves in Israel.[14, 15] Establishing when the habitual use of fire emerged is of major importance for human evolutionary history, but again the issue comes down to what constitutes a 'controlled' use of fire.[16]

A good example of the problems encountered when identifying the first use of fire comes from the evidence that has been derived from the site in England called Beeches Pit, at West Stow in Suffolk. This is an interesting Lower Palaeolithic site that dates from around 400,000 years ago, and evidence of burning comes from two distinctive levels. The site occurs within alluvial interglacial sediments.[17] An immediate problem is that charcoal derived from natural wildfires may occur naturally in fluvial sediments. Charcoal may also be reworked. So how can we establish whether the charcoal at this site was related to human activity? As we noted earlier, in order to be certain, we need to determine whether the evidence of burning is localized in the form of a hearth; whether there is evidence of human activity on the site, such as burned flint flakes; and whether there is evidence of the effect of heat from the fire on the surroundings. In this case the lines of evidence appear clear, and it has been suggested that humans sat around the fire knapping flints, though interpretation of temperature data from the study of charred and calcined bones in the hearth remains more controversial.

Even in a site such as Beeches Pit there is still much more that can be done. The identity of charcoal associated with such sites may be useful, as would temperature data derived from charcoal reflectance. We should compare charcoal data from a variety of hearths, both from archaeological sites and from experimental work, and look at these against charcoal from wildfire-derived charcoal sites.

It is probable that the first use of fires was to keep warm at night and as a source of light in a darkened cave, but humans soon also controlled fire for cooking. Richard Wrangham has looked at the archaeological evidence of the use of fire for cooking. It is possible that fire may have been used in this way some 1 million years ago, as there is evidence of burned bone that may have been part of the meal that was cooked.[18] There seems to have been increasing potential for the use of fire for cooking through the middle Palaeolithic, 300,000–50,000 years ago.[19]At any rate, many archaeologists suggest that cooking was not a regular activity until the late Palaeolithic (that is, from 50,000 to as late as 10,000 years ago).[20]

Our understanding of and evidence for the human use of fire is fraught with problems. Not least of these is one of intent. There is also the difference between the use of fire and control of fire. This problem is multifaceted. Not only do we need to demonstrate the close association of fire and human activity, but we need to distinguish between 'fire users' and 'fire producers'. When did humans change from being fire users or keepers to becoming fire producers and managers? This switch is very difficult to date. While we have evidence of humans and fire perhaps as far back as 1.5 million years ago, fire production (for example, the use of flints to produce a spark) may not have occurred until the very end of the Pleistocene, perhaps even as recently as 40,000 years ago (Figure 55).

How to obtain data on, or interpret the evidence of, the changing use of fire during this early period presents a challenge. As Andrew Sorensen and colleagues have put it, 'we archaeologists have yet to ascertain, even in coarse chronological terms—when in our early prehistory fire became a standard part of the human tool kit'.[21] Clearly there is significant potential for experimental archaeology here to help us understand what evidence we should be looking for in the fossil record.

Chris Stringer notes that cooking was an important milestone for hominins in terms of making meat more digestible and neutralizing pathogens and toxins.[22] And of course it also has a social role. Yet there is a significant difference between cooking meat and the use of cereal crops that are only edible after cooking. Evidence that Neanderthals also used fire to cook comes from Shanidar Cave in Iraq and Spy Cave in Belgium. Phytoliths—silica deposits found in many grasses, including cereals—have been found here, as well as starch grains from dental tartar showing distinctive markers of cooked plant food. Such complex tasks are likely only accomplished in a social context.[23] Humans took a further step. Grains could be planted, harvested, and stored. The development of the preparation and cooking of grains led to a major change from a hunter-gatherer way of life to a settled, agricultural one.

Some important aspects of the growing of grain and its spread have come from charcoalified grains. These were stored in large areas but were subject to fire, and hence preserved as charcoal. I first became aware of this important research through the Ancient Biomolecules Initiative, a collaborative research programme run by the UK's Natural Environmental Research Council.[24] Cambridge and Manchester scientists were able to extract DNA information from charred grains of ancient cereals.[25] New techniques have expanded the possibilities of this approach, and enabled the tracking of the geographical spread of cereal varieties in prehistory.[26] In particular, the research has been able to distinguish between wild varieties and domesticated forms. The spread of agriculture from the Middle East has been an area of particular interest. A team from the University of Manchester has shown that the domestication of crops originally took place in western Asia, and that they were introduced into south-eastern Europe around 7000 BC. The domesticated crops then spread across Europe, and charred grains from around 5000 BC have recently been discovered in Spain.[27]

Similarly, charred grape seeds have allowed us to trace the development of viticulture and winemaking in some ancient cultures.[28] The charred seeds and sometimes the grape skins provide sufficient information to allow their identification. It may seem surprising that, even in charcoalified plant assemblages, pressed grapes could be identified, and distinguished from whole grapes or raisins.

In the context of excavated early urban centres, fire is usually seen as destructive. Residues of charcoal in archaeological sites are often used to identify layers of destruction, either accidental or deliberate. But charcoal may be introduced into an archaeological site through natural processes too. We have already seen that it can float for long distances and is also very resistant, so it is necessary to establish the origin of any charcoal. As charcoal contains anatomical data, it may be possible to demonstrate that it comes from destroyed building timbers rather than the burning of the surrounding vegetation (though dating the fire may be difficult, as carbon dating may pick out the age of the growing wood). Recent research on growth rings in charcoal and the reconstruction of the diameters of the trees or branches that were burned can help here, as the trunk diameters of charcoalified wood from an urban context should be uniformly large, while that from a natural wildfire setting may show a wide range of diameters.[29]

When we consider fire within an archaeological and urban context there is a danger that we might forget about fire in the landscape. It is quite possible that natural wildfires may have spread into an urban area and destroyed a house, village, town, or even a small city. We can see that today, when fires in North America or Australia have spread to habitations causing, in some cases, total destruction of communities. In such cases, if there wasn't a study of natural wildfire signals from outside the urban context, I wonder if archaeologists of the future would see the destruction as caused by a purposeful human act.

The environmental impact of human fire use

Agriculture may have originated by accident rather than planning, but once it had become established, fire is likely to have been used as a key agent to clear the land, a role that it still fulfils today. The relationship between fire and hunting is much more difficult to prove, but is a topic of intense interest and debate. Pyrodiversity has been defined as the outcome of complex interactions between fire regime, biodiversity, and ecosystem effects. There is mounting evidence that burning by indigenous peoples has played a significant role in shaping pyrodiversity as well as affecting plant and animal community structure. This practice also appears to reduce the incidence of large, lightning-started fires.

We can try to gain insights into early human fire use by examining cultures where the practice is still undertaken today, such as in parts of Australia and Africa. For example, the indigenous peoples of Australia (Aboriginal Australians) practise a fire-stick culture, where they use fire to drive game into an area to be killed.[30] This practice is also used in Africa.[31] The burning of vegetation can, following rains, give rise to luxurious growth that attracts game animals.[32]

In North America it appears that the peoples who colonized the continent used fire both for clearing the landscape and also for hunting, much as in Australia.[33] Fires appear to have been set by the different Native American populations to target bison, deer, and antelope, and to drive buffalo to their death. Across the world, as Stephen Pyne points out, 'any creature that could be hunted by fire was'. This included using torches to help fishing at night, smoke to flush out bears from their dens, and fire drives to flush out springbok in Africa and kangaroos in Australia.

Fire has been a powerful agent of biotic forcing. Not only can fire be used to change or even promote different kinds of plants, but also to encourage better crops, as in the case of acorns and chestnuts. These trees are more tolerant to fire than many others, and in both cases it has been found that seed production significantly increases following a fire when the tree has been affected but not killed, because the tree invests unused food that is in storage to produce an enhanced seed crop the following year. Fire can also be used to drive off insect pests, both from crops and also away from humans.

Fire can control grasses, creating conditions for a new flush of growth in the correct season, and even for a second flush. The use of fire to control woodland development is one with which we are grappling today, but Native Americans had for centuries set surface fires in the pine forests of south-western North America, and this meant that while there were regular fires, there were few, if any, crown or megafires.[34]

One of the classic pieces of evidence of the human use of fire is that from hearths and pits, but, as we have noted, it is not always easy to identify definitive hearths or to demonstrate that human activity was involved. Recognizing cases of human activity as the major cause of fire on the landscape is much harder still. In some cases, evidence for increase in fire activity, such as in Australia, has been used to argue for the arrival of humans into the continent.[35] But it is equally possible that changes in fire regime may have been the result of changing climate and/or vegetation, and in some cases even changes in the preservation of charcoal. As we have seen, there are many reasons for the occurrence of charcoal in the fossil record. Debates continue to rage concerning the impact that human activity has had upon fire in the landscape, and to what extent fire records can be used to indicate human activities.

Another approach to identifying human-initiated fire impact on the environment is to use charcoal data from deep-sea cores. We have already seen how such data were used to chart the increase in biomass burning through the later Cenozoic, when grasslands began to burn. It has been suggested, using marine core data, that there is no evidence of extensive fire use for eco-system management by Neanderthal and Upper Palaeolithic modern human populations, and that the fire data of burning between 70,000 and 10,000 years ago follow natural climate variability.[36] Human fire, it has been argued, did not have any influence on a regional scale. However, what this type of analysis cannot show is the timing of fires. Not only do humans change fire by adding or taking fire from the landscape, but more par-ticularly they can change the period of burning, often burning earlier in the season or at the end of the normal fire season. Such changes would not be observable using charcoal data.

Since we humans started our journey with fire, we have found many new ways to use it. The simple use of fire for heat and for cooking has been joined by the use of fire to manipulate the environment, from slash-and-burn agriculture to full landscape transformation. However, in all these changes we should not lose sight of the fact that natural fire regimes are affected by vegeta-tional change, and also fundamentally by changes to the climate.

As mentioned in Chapter 6, I had become involved in a project to unravel the fire history of the California Channel Islands because they were the site of the oldest human remains in North America, and we could compare patterns of fire before and after human arrival. What became apparent from our research was that there was evidence of landscape fire in coniferous forests from at least 24,000 years ago, some time before the arrival of humans around 12,000–13,000 years ago. We found evidence of significant burning through mixed forest from 13,500 to 12,000 years ago,[37] and we may be tempted to ascribe this burning to humans clearing the landscape—or even perhaps using fire to hunt the small mammoths that lived on the islands.[38] But there is another factor to consider. This was a period of dramatic climate change, with a major climate reversal some 12,900 years ago with the onset of the Younger Dryas cool interval. The Younger Dryas saw a brief reversal of the warming trend following the end of the last major glaciation. Many believe that this may have been caused by ice melting and a sudden input of fresh water into the North Atlantic that affected the Gulf Stream. We might well expect both vegetation and fire regimes to also change in response, so unravelling these changes from human activity is very difficult indeed.

Fire and climate

As we have seen, there is ample evidence from the charcoal record that fire and climate are interlinked. It has often been argued that human activity may be the cause of many fires, but a fire will not take hold and spread if the fuel loads and weather conditions are not right in the first place. We can reduce the potential for human ignitions, but this will not lead to the exclusion of landscape fire.

There is increasing evidence that is widely and overwhelmingly accepted that human activity is accelerating climate change through the generation of CO_2 and other greenhouse gases. While it is clear that CO_2 emissions have been rising over the past decades, the contribution of biomass-burning emissions as opposed to fossil fuel emissions has been difficult to untangle. This is significant, as both play a role in global climate forcing, and satellite-derived data have been used to separate the two. The distribution of CO_2 emissions from biomass burning and fossil fuel combustion obtained from satellite data has shown that combustion practices may be shifting from open landscape burning to contained combustion for industrial purposes.[39]

At the beginning of the book I mentioned Stephen Pyne's speculation that as the world's population moves from the countryside to the cities there is a 'pyric transition', whereby agricultural fires decline, wildland fire is suppressed, and fossil fuel use increases. This change also has a psychological dimension, with fire management and use supplanted by ideas of fire suppression and even eradication. As we have seen, fire suppression causes the build-up of surface fuels, so that subsequent fires become more intense and potentially more devastating. The USA woke up to this fact following the fires in Yellowstone National Park in 1988.

Western US Forest Wildfires versus March to August Temperature

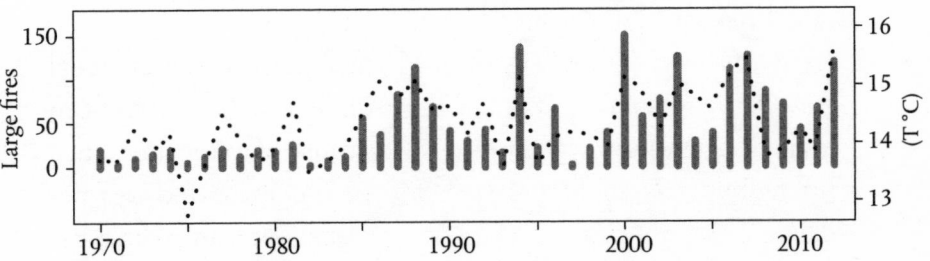

Figure 57. Annual frequency of large (greater than 400 hectares) western USA forest fires (bars) and mean March–August temperatures (dotted line). Since 1980 there has been an increase in the number of large fires during more frequent periods of higher mean temperatures.

Changes in climate are now having an impact on fire activity, for example in the western United States (Figure 57).[40] Climate change may allow the spread of new vegetation types, altering fire regimes, and the spread of insect pests that have caused tree death, increasing dead fuel loads, and again promoting an increase in fire activity.

There has been increasing debate too on the use of prescribed burning, as well as, in some cases, a 'let the fire burn' policy, but as with all problems the causes and solutions are not simple. In many places today we have an artificial landscape, which is coping not only with changes in climate but also changes in vegetation, as a result of climate change and also because of introduced species. To this we may add two further elements—the desire to build into wildland environments, and also the desire to protect not only life but also property.

As settlements increasingly encroach on wildlands, the fact is that humans, even if they are not actively managing fire in the landscape, have to understand fire in order to have a sensible policy. A balanced approach is difficult. How can we let a flammable landscape burn when people and habitation centres are threatened? To what extent should we actively manage fire with prescribed burns, when smoke from the fires may also have a health consequence for some?

8

The future of fire

Some say the world will end in fire,
Some say in ice.
From what I've tasted of desire
I hold with those who favor fire.

Robert Frost, 'Fire and Ice'

I only encountered the term 'wildland–urban interface' a few years ago. It describes situations or physical boundaries where human urban populations and infrastructure impinge on wild vegetated areas. Two specific cases are worth highlighting. One is simply due to the expansion of population centres, where towns and cities continue to spread into rural areas and, in some cases, impinge on natural vegetation. The other situation occurs when individuals or small communities build homes and infrastructure within the bounds of an area of wild vegetation. The ultimate getting away from it all! This wish for exclusivity and privacy is growing at an ever-increasing rate and is becoming a major global challenge.[1] And even before the houses and communities encroach into the wilderness, the natural vegetation is experiencing the effects of human activity and climate change.

The impact of invasive plants

Simply put, an invasive plant is a plant that has gone wild in an area where it never occurred naturally before being introduced. We are all familiar with bringing exotic plants into our garden, but less aware of what happens to the plants if they spread outside our own area. In general this may not be a problem. Across many parts of the world, introduced plants are confused with natives. Rhododendrons, for example, are very widespread in the UK, and in some places they may also be considered a 'weed'. But they were introduced into our gardens from China. In any case, what does it really mean to say a plant is native? It isn't always obvious. While the cultivated species *Rhododendron* may have been a relatively recent import to Britain, wild forms did exist in England more than 55 million years ago.[2] Equally, we may not realize that a plant is not a native of a region, or what potential problems they may cause. While in some cases such plants may be escapees from our gardens, plants may also have been introduced for another use, such as to provide feedstock for animals. There are those who think that plant invasives are not really a problem, but I would challenge this view in relation to fire.[3]

In some instances, plants that were introduced for very good economic reasons have turned out to have unintended consequences. One of these is the *Eucalyptus*. This tree, as we have seen, has many fire traits. It has evolved in a fiery landscape.[4] Indeed, it is designed to burn. However, we are now finding that plantations of these trees in other areas have altered local fire systems, for instance in Portugal, where fires that start in the eucalypt plantations during dry spells become more severe and intense, and can spread into native vegetation with disastrous consequences, as was seen in the fires of May 2017 where so many were tragically killed. It is only in the past few years that an understanding of flammable vegetation and its long geological history has begun to inform our debates on this particular problem.

In some cases an invasive plant is simply a nuisance, in other cases it can suffocate and replace native species and be difficult to eradicate, such as Japanese knotweed. When it comes to fire, it is the invasive grasses that are causing the most problems. In most cases the grass has been introduced as animal feed. For example in Australia it is gamba grass that has proved to be the biggest problem. It was introduced because it grows fast and provides a good source of food for cattle. But the grass has escaped, with disastrous consequences because of its flammability. Gamba grass grows tall, and quickly, so any fire in the grass burns much hotter and more fiercely than normal surface fires. This has major implications for the survival of native vegetation that copes only with lower-temperature surface fires.

A similar problem exists in North America with cheatgrass (*Bromus tectorum*). Here the grass has spread across a wide range of habitats and is visibly altering fire regimes.[5] In some cases it provides a surface fuel that causes a fire to spread more readily. In many cases the grass spreads along the side of highways and provides a ready conduit for fire. The grasses may also thrive in drier environments, and their spread is enhanced by climate change. We have already discussed the natural grass–fire cycle, but the spread of cheatgrass is bringing this interaction into areas of native vegetation. In some cases the spread of cheatgrass has been catastrophic. The iconic saguaro cacti (*Carnegiea gigantea*) seen in most Hollywood westerns are found in the western USA (Figure 58). While these cacti may be struck by lightning, they are usually isolated, so that fire will not spread between them.[6] But with the spread of invasive grasses, the fire can spread from one cactus to another and destroy the whole ecosystem. Unless the grasses are eradicated, the saguaro ecosystem is unlikely to survive another 30 years.

Figure 58. Invasive grasses spread between saguaro cacti in Sonoran desert, USA.

The interrelated problems of fire, plant invasives, and climate change have only become apparent from the study of recent data derived from satellites. However, yet another problem has arisen from the attempt to limit the impact of fire in some areas. We have already noted the effects of post-fire erosion. The forestry industry has been under pressure from two fronts. The first is to lessen the impact of run-off from recently burned areas. One such attempt was to drop bales of straw from helicopters or air-craft to help cover the bare soil and to absorb rain so that there was no simple run-off. But some of the straw bales that have been used in the past have contained seeds of exotic grasses, accidentally contributing to the spread of the invasive grasses.

The second issue facing the forestry industry is the requirement in some places to replant native vegetation following a fire—to revert the area 'back to nature'. Yet by 'natural' is meant a habitat that has been managed by humans for hundreds or even thousands of years, and significant climate change will alter the distribution of plants naturally anyway. Only a long-term and historical perspective may help in such circumstances.

Fire in the warming Earth

We have already seen the influence of climate on fire regimes through Earth history, and in looking to the future, we must consider fire in the context of climate change. We cannot avoid the fact that slash-and-burn agriculture is still a fact of life in many parts of the world. For some populations this is the only way to survive.[7] The problem is particularly acute in places such as the Amazon rainforest, where important ecological resources are being destroyed (Figure 59).[8] This is all part of the process of land clearance and changes to land use. We need to find a middle way

Figure 59. Slash-and-burn deforestation fire in the Amazon rainforest in 2006, Mato Grosso, Brazil.

that allows the use of fire management by indigenous populations and separates the activity of commercial loggers from
those who have traditionally used the landscape for survival.[9]
The problem is not confined to areas of natural vegetation, but
also arises in the context of river flood plains, where excessive
building makes flooding worse. This can be coupled with the
threat of fire and post-fire erosion, leading to ever-more damaging
fire–flood cycles. We have already seen how the problem of fire
and floods can be interlinked. In Colorado, for example, there
were a number of significant fires in 2011 both north and south of
Denver, including the Boulder area.[10] There is no doubt that subsequent flooding in 2013 occurred partly as a consequence of the
fires in the previous years.[11]

Our increasing scientific understanding of fire in recent years
has not translated into greater public awareness. It is hard to argue
against the perception that the freedom to build in a flammable
landscape is an individual's 'right', even though building in such
areas may put not only their own lives at risk but the lives of those
tasked with extinguishing fires. The deaths of firefighters are
increasingly causing questions to be raised as to whether a fire,
even one threatening damage to property, should always be extinguished.[12] I wonder if we will get to a situation of 'let the builder
beware'—if you build in a flammable landscape then there will
be no help in putting the inevitable fire out. But no politician is
likely ever to go that far, as there would always be those who would
stay to protect their property and risk death in the process.[13]

If we cannot always prevent fire in some areas we may be able,
at least in part, to help reduce the causes of human-started fires.[14]
It is difficult to prevent arson completely, but the requirements to
restrict the use of campfires, barbecues, and even smoking (dis-
carded cigarette ends can also be the cause of a fire), may be more
vigorously enforced. A significant problem is that it becomes

difficult to explain why natural fire in some types of vegetation and environments is a good thing, yet in other places and types of vegetation it is bad.[15]

Nowhere is this debate more difficult than in the area of prescribed burning, as a way of managing fire on the landscape. In some cases, prescribed burns have got out of hand and transformed into a full-blown wildfire.[16] While prescribed burning may seem a useful tool in the forester's armoury, there are additional issues that need to be understood when looking at forest management and fire.[17]

In some cases the impact of human actions may be clear, as in the case of the peat fires in Indonesia. Draining of the peat areas has led to increasing fire risk (Colour Plate 14). This is also partly done to allow logging of the forest trees, which in turn has consequences for fire in an environment where fires are not normally seen. The situation may be made worse by climate cycles that can have a significant impact on the newly drained areas. During El Niño years the climate is distinctly drier in areas of Indonesia, so that fires that now occur partly as a result of human activities may become much more serious as El Niño events intensify as a result of climate change. Between September 1982 and July 1983, fires spread on the island of Borneo over an area of over 37,000 km² —equivalent to that of Belgium and Luxemburg combined— and it is now believed that the size of the fire was a direct result of the fact it occurred in an El Niño year.[18]

Fire and forestry policy may also have unintended consequences. In Russia there is a moratorium on logging in some forestry areas, for example in parts of Siberia. However, loggers are allowed to remove trees killed by wildfire. This poses a temptation, and there is some evidence that during years where there are natural fires in such forest areas, other fires may have been started intentionally to allow a much wider area to be logged.

It has become increasingly clear over the past decade or so that there are a number of problems concerning fire with regard to ecological conservation. We now know that fire is an important factor for the health of many types of vegetation worldwide.[19] Yet it is hard to overcome the perception that it is always bad, and that it can have serious consequences. In some cases the desire to protect an area by the exclusion of fire may, in the long run, put the area that we wish to conserve at risk. This is a dilemma in places like California, where there are major flammable ecosystems. Chaparral vegetation is a case in point (Figure 60). There is no doubt that fire is a natural part of the chaparral ecosystem, but when, therefore, should fires be allowed to burn, and when extinguished in such circumstances? A misunderstanding of the nature of fire in such areas may lead to incorrect policies being pursued by politicians and conservationists alike.[20]

Another problem is compounded by several misconceptions. Many consider grasslands, especially in Africa, to represent a degraded landscape and a result of the destruction of forest. The implication is, therefore, that we should convert these grasslands to forest by planting trees. Yet it has been shown that these grasslands are not only ancient but rich in plant species, many of which rely on fire for their survival.[21] We are therefore faced with situations in some parts of the world, such as some areas of Madagascar, in which not only is fire necessary, but planting trees is not helpful in maintaining biodiversity. As we saw in Chapter 1, in Madagascar fire is not good for some ecosystems yet beneficial for others, and a 'one-size-fits-all' policy for fire may be a mistake.

Reasoned debate concerning fire can be difficult when there are interest groups with very different points of view. Nowhere is this more evident than in the UK with regard to the burning of heathlands. Some groups favour the total exclusion of fire, while

Figure 60. (a) Chaparral vegetation in southern California; (b) burnt area.

others advocate a managed burning programme. The burning of heathland plays a role in maintaining the ecosystem, but smoke from this burning may be hazardous. And what may be bad for some animals may be beneficial for some plants. A lot depends on whether there is a specific consideration, for example the effect on bird life, or whether we are assessing the impact on the ecosystem as a whole. This is a complex issue with no completely right or wrong answer.[22] And the considerations need to be weighed up against a backdrop of changing climate, which will itself be altering the vegetation.

Fire as health hazard

Health concerns related to fire have only come to prominence in the past few years.[23] The problem is not so much the fire itself but the smoke derived from the fire. Smoke from fire can spread a very long distance, in some cases many hundreds or thousands of miles from the fire itself. For many, fire smoke is nothing more than a temporary nuisance. But when we are dealing with large fires the smoke can linger for days and cause a number of problems for human populations.[24] Smoke inhalation can cause breathing difficulties not only for those who are asthmatic but even for those without any lung problems, and in some cases can lead to death. We have already seen that smoke from Indonesian fires can spread many hundreds or indeed thousands of miles across Asia. The global distribution of smoke-affected deaths shows concentrations in regions where there are major and frequent fires.[25]

Breathing problems may not be the only effect of fire on health. Pregnant women can be badly affected by wildfire, and this is giving rise to concern over the number of premature births and birth abnormalities that may be linked to exposure to smoke at certain periods during pregnancy. As we continue to extend towns into wildland areas, and expose populations to frequent smoke from fires, many of these issues are likely to get worse, not better.

How can we be certain that fire has been affected and will in the future be affected by climate change? Over the past few years it has become possible to link aspects of fire and climate more precisely so that a number of different issues may be identified. This is not only because of a greater understanding of climate, but also due to improving records of fires on a regional, continent-wide, and even a global basis. It is possible to demonstrate whether there has been more or less fire on a landscape, and also identify when the fires occurred. This is significant, as the timing of a fire may have its own ecological consequences. We may be familiar with fire seasons, such as in southern California, but if those seasons are changing this may have implications for future fire policy.[26] If we look at fire in Africa, we can see that fires burn in different parts for very different reasons. The fires in central Africa burn at a different time from those in southern Africa. However, many fires in central Africa are human-ignited fires, whereas many in southern Africa are natural, lightning-ignited fires. There is currently debate over what counts as natural fires in natural vegetation, as opposed to managed fires in such areas. Only recently has there been a realization that some of the grasslands are ancient, and shaped by natural fire, whereas in other areas the grasslands are newly introduced. Unravelling the complex history of vegetation and fire in Africa is an important challenge, as misunderstandings can lead to incorrect conservation efforts, let alone political fire-management decisions.[27]

We also know from our growing understanding of fire through time that increasing temperatures can lead to increasing fires.

Not only that, but changes in the distribution of rainfall may have a significant effect, from an early wet season giving rise to increased plant growth to a later dry season resulting in more extensive and perhaps severe fires.

And we should not forget that subtle changes in climate might allow fire to become part of a landscape where in the past it was absent. In other words, areas where lightning starts fires may shift as the climate changes. Our fire policies need, therefore, to be continually revised in the light of this.

For me, personally, this is not simply an academic question. I live in Surrey, in the United Kingdom. This is a highly populated county in the south-east of England, and not an area famous for its wildfires. Surrey is, however, one of the most forested areas in England, and a change in the nature of fire regimes as a result of climate change may have significant consequences. Heathland fires occur here on a regular basis, but although there may be some forest fires these are generally surface fires that have been controlled. Imagine the havoc if any of these fires should become major crown fires—not only would we have difficulty in extinguishing them, but the possible consequences to housing, people, industry, transport, etc. would be catastrophic. In a situation like this it is important that contingency planning is made. But how can we know if an area where fire has not been the norm might change? Can we plan ahead?[28]

In recent years growing effort has gone into forward climate and fire modelling. While climate and vegetation models looking at future climate change scenarios are well known, the integration of fire into such models is still in its infancy. There are two major approaches to the problem—one is based on historic data, and the other uses fundamental principles of how the Earth functions. Whichever approach is favoured, they allow us to predict where and when major changes in the fire system may occur, and hopefully alert governments to the need to alter their attitude towards fire.

One thing is clear: if we are to cope with fire in our present and future world, we need to understand the 400-million-year history of fire on Earth, its role in the Earth system, and our fundamental relationship with one of nature's most powerful forces (Figure 61).

To tackle the issue of fire in a world that is undergoing climate change may need some new approaches. In a recent meeting at the Royal Society in London, some 30 of the world's leading fire scientists signed up to what is known as the Chicheley Declaration.[29] This sets out the dangers to the increasing population of a deeply interconnected world of changing patterns of fire as the effects of climate change set in. It appeals for greater research into wildfire, more communication and public debate, and an integrated, multidisciplinary approach to solving the challenges posed by future fires.

So what have we learned from our 400-million-year journey, and why is it important? It is evident that fire has been a significant force on the Earth for millions of years, long before humans appeared on the scene. Fire plays a role in the regulation of atmospheric oxygen, which we all need to survive. Many plants and animals evolved in fiery landscapes and not only adapted to fire, but in some cases need fire for their survival and spread. But we humans have adapted fire to our own needs, and in the past we had learned to live with fire. Fire, then, is an integral cog in the way in which the Earth works. Is it right or desirable to remove fire from the landscape? Clearly if we are interested in conserving many of the world's richest environments this is not an option. If that is the case, we have to learn to live with fire. The consequences of this may be unpalatable to some, and may mean there are changes to policies relating to developments in some wild

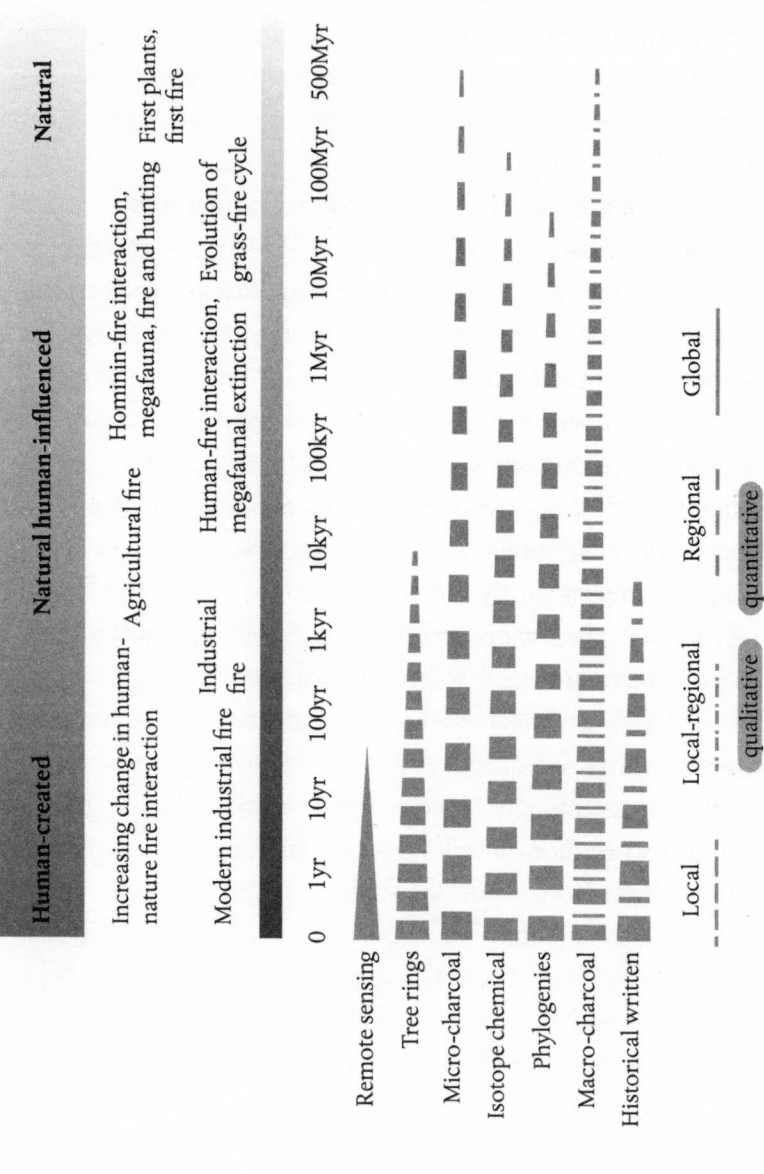

Figure 61. The evolution of natural and man-induced fires and methods of study.

areas as a result of allowing fires to burn even if it means the destruction of property.

In the final reckoning, the decision may be out of our hands. We may imagine we can control fire, but in many instances this is an illusion. One of the greatest challenges over the next decades will be the availability of water, and the driest areas where water is scarce are also likely to be prone to wildfire, so the use of water to extinguish a fire may be seen as a luxury.[30]

We need to be prepared for fire in the future. We need to understand how the changing water supply will impact on fire and how we engage with it. We also need to remember that fire sees no international boundaries, and may have the potential to create crises and conflicts. As a global population we need to re-engage with fire, and recognize our evolutionary history of living in a fiery world. With climate and vegetational changes, fire may become a problem where it was not in the recent historical past. We can no longer think that if we have never experienced a wildfire in our lifetime then we will never have to face that possibility.[31] This means that we need to increase education about fire—not only the negative but also the positive aspects of fire.[32] To face the future, we must recognize that fire is a natural and important part of how the planet works, and that we have much to learn from the 400-million-year history of fire on Earth.

APPENDIX: INTERNATIONAL GEOLOGICAL TIME SCALE

Era	Age (Myr)	Period	Epoch
Cenozoic	2.6 23 66	QUATERNARY NEOGENE PALEOGENE	Pliocene Miocene Oligocene Eocene Paleocene
Mesozoic	145 201 252	CRETACEOUS JURASSIC TRIASSIC	Upper Lower Upper Middle Lower Upper Lower
Paleozoic — Early	299 359 419	PERMIAN CARBONIFEROUS DEVONIAN SILURIAN	Longpingian Guadalupian Csuralianpian Pennsylvanian Mississippian Upper Middle Lower Pridoli Wenlock
Paleozoic — Late	444 485 541	ORDOVICIAN CAMBRIAN	Ludlow Llandovery

Appendix. Geological Timescale showing absolute ages in millions of years. Ages based on the 2017 International Chronostratigraphic chart produced by the International Commission on Stratigraphy. <http://www.stratigraphy.org/index.php/ics-chart-timescale>.

GLOSSARY

Angiosperm A vascular land plant that reproduces using flowers and produces enclosed seeds. Ovules are enclosed within an ovary and when pollinated the seeds develop within a fruit that has developed from carpels.

Bogen-structure Shattered cell walls mainly in fusinite as seen under the microscope in reflected light. The shattering takes place after the plant material has been buried and the shattering indicates that it was brittle before burial, such as in charcoal.

Bone bed A rock layer that has an unusual concentration of vertebrate bones.

Cellulose The building block of plant cell walls. An organic compound comprising carbon, hydrogen, and oxygen comprising linear chains of linked D-glucose units.

Crown fire A fire that spreads into the upper storey or crowns of trees. Fires usually start as surface fires but may spread via ladder fuels up trees into their canopies (or crowns).

Fusain Fossil charcoal. Fusain shows high reflectance and dull lustre in different angles of light, and is also characterized by its chemical inertness, being almost pure carbon. The term was introduced by Marie Stopes as one of the four fundamental constituents of coal (vitrain, clarain, durain, and fusain). She chose the name fusain (the French term for charcoal) as the label for that charcoal-like component of bituminous (black) coal.

Gymnosperm A vascular plant that reproduces using naked seeds. Many bear their seeds in cones. Typical modern groups include the conifers and cycads.

Horizon A distinct layer of rock or soil.

Inertinite A maceral of coal that has high reflectance and shows anatomical structure. The inertinite group macerals include fusinite, semi-fusinite, inertodetrinite, macrinite, micrinite, funginite, and secretinite, Inertinite is a maceral group that comprises macerals whose reflectance in low- and medium-rank coals, and in sedimentary rocks of corresponding rank is higher in comparison to the macerals of the vitrinite and liptinite groups.

Isotopes Atoms of an element with the normal number of protons and electrons, but different numbers of neutrons. Isotopes have the same atomic number, but different mass numbers (number of protons and neutrons). Oxygen for example may have a mass of 16 or 18. Carbon may have 6, 7, or 8 neutrons and 6 protons. Both C12 and C13 are stable but C14 decays and is radioactive.

Isotope excursion This is where the values of a chemical isotope (such as carbon or oxygen) may depart in a negative or positive direction from the background in a vertical sequence of data points through the rock record. Such excursions indicate some perturbation in the Earth's atmosphere and may be global in nature.

Lamination Small-scale sequence of fine layers (laminae). A distinct layer of sedimentary rock that is on a smaller scale than a bed or horizon.

Lignin The chemical important for strengthening plant cell walls (usually up to 30 per cent in wood). An organic polymer comprising carbon, hydrogen, and oxygen comprising aromatic rings. These are composed of three different phenyl propane monomers. Lignins differ in several plant groups; conifers and angiosperms have different compositions.

Lignite A type of coal. Lignite was initially formed as peat. During burial it underwent alteration, mainly through being subject to an increase in temperature (and some pressure), and reached a stage in the coal series known as the lignite rank. It may have 60–70 per cent carbon but also a high moisture content. The rank series indicating progressive alteration, or coalification, is peat to lignite (also known as brown coal) to subbituminous to bituminous coal (also known as black or hard coal) to anthracite.

Maceral The basic language of coal petrology concerns coal macerals. The term maceral was first introduced by Marie Stopes in 1935 who proposed the word (from the Latin *macerare*, to macerate) as a distinctive and comprehensive word tallying with the word 'mineral'. Macerals are organic substance, or optically homogenous aggregates of organic substances, possessing distinctive physical and chemical properties, and occurring naturally in the sedimentary, metamorphic, and igneous rocks of the earth.

The **Paleocene–Eocene Thermal Maximum (PETM)** was a short-lived rapid increase in global warming (increasing between 5°C and 8°C) that took place around 55.5 million years ago and lasted around 200,000 years. The episode is marked by a large carbon isotope excursion.

Petrography (as branch of petrology) Concerns the description and classification of rocks (often aided by microscopic examination).

Post-fire erosion The impact of surface fires on the landscape may be severe. Fire may not only destroy plants but also rooting systems that bind the soil. Fire may also make the soil hydrophobic (resist water). Rainfall following fire may cause sudden loss of soil and transport of sediment by overland flow. Deep gullying may also result. This is termed post-fire erosion.

Reflectance Where light is reflected from a surface. In microscopic studies of organic material light is reflected and observed from polished surfaces of the specimen, usually under oil. Measurement of the reflectance provides data on the temperature history of the material.

Scanning electron microscope (SEM) This is a type of electron microscope that produces high-magnification (many thousands of times) three-dimensional images of specimens (with a resolution better than 1 nm). The image is produced by scanning the specimen with a focused beam of electrons, usually in a vacuum. The most common way of obtaining an image is by recording the secondary electrons that are knocked from the surface of the specimen. A three-dimensional image is produced. The images produced are black and white but some may be artificially coloured.

Stomata (singular stoma) Openings or pores that are used by plants for gas exchange and are usually found on the surface of plant leaves. Air enters through the openings, photosynthesis converts the carbon dioxide to sugars, and oxygen is expelled as a by-product. The stomata are opened and closed by guard cells that also control water loss from the plant (transpiration).

Surface fire A fire that burns both dead fuels and living plants on the forest floor. They also occur in grassland and scrublands.

Vascular land plant A plant that has vascular tissues used for water and food transport. These tissues include both xylem (water-conducting cells), which are often lignified, and phloem (food) transport cells. Vascular plants are also called higher plants or tracheophytes. The most common modern types are ferns, club mosses, horsetails, gymnosperms, and angiosperms.

Xylem A transport tissue used in plants. It generally comprises tracheids (thin elongate cells), but in some flowering plants larger vessels occur. Xylem is used to transport water (and some nutrients) from the roots of the plant to the shoots and leaves. Other cell types such as parenchyma and fibres also occur. Secondary xylem is known as wood, and the cellulosic cell walls are strengthened by a resistant chemical known as lignin (a carbon compound with an aromatic or ring structure).

NOTES

CHAPTER 1

1. Kull, C.A. (2004). *Isle of Fire: The Political Ecology of Landscape Burning in Madagascar*. University of Chicago Press, Chicago, IL.
2. TV documentary. Many documentaries were shown on British TV in the late 1990s and early 2000s. One on ITV in 2002 called *Anatomy of Disaster* concerned the Californian fires. It was made in America in 1997 by grbtv.com.
3. Roy, D.P., Boschetti, L., Smith, A.M.S. (2013). Satellite remote sensing of fires. In: Belcher, C.M. (ed.). *Fire Phenomena and the Earth System: An Interdisciplinary Guide to Fire Science*, 1st edition, pp. 97–124. J. Wiley & Sons, Chichester.
4. <http://www.firelab.org/project/farsite>.
5. There is an excellent account of the fire and aftermath, the summary of which is freely available from the Internet. Graham, R.T., Technical Editor (2003). *Hayman Fire Case Study*. Gen. Tech. Rep. RMRSGTR-114. Ogden, UT: US Department of Agriculture, Forest Service, Rocky Mountain Research Station.
6. For a detailed discussion of fires, types, and spread see Scott, A.C., Bowman, D.M.J.S., Bond, W.J., Pyne, S.J., and Alexander M. (2014). *Fire on Earth: An Introduction*. John Wiley and Sons, Chichester.
7. <https://en.wikipedia.org/wiki/Rim_Fire>.
8. Moody, J.A. and Martin, D.A. (2009). Forest fire effects on geomorphic processes. In: A. Cerda and P. Robichaud (eds), *Fire Effects on Soils and Restoration Strategies*, pp. 41–79. Science Publishers, Enfield, NH; Moody, J.A. and Martin, D.A. (2001). Initial hydrologic and geomorphic response following a wildfire in the Colorado Front Range. *Earth Surface Processes and Landforms* 26, 1049–70.
9. Meyer, G.A. and Pierce, J.L. (2003). Climatic controls on fire-induced sediment pulses in Yellowstone National Park and central Idaho: a long term perspective. *Forest Ecology and Management* 178, 89–104.
10. <https://en.wikipedia.org/wiki/Yarnell_Hill_Fire>.

11. Johnston, F.H., Henderson, S.B., Chen, Y., Randerson, J.T., Marlier, M., DeFries, R.S., Kinney, P., Bowman, D.M.J.S., and Brauer, M. (2012). Estimated global mortality attributable to smoke from landscape fires. *Environmental Health Perspectives* 120, 695–701.

12. See picture in Scott et al. (2014), fig. 1.44, p. 46.

13. Holloway, M. (2000). Uncontrolled: the Los Alamos blaze exposes the missing science of forest management. *Scientific American* 283, 16–17.

14. Peluso, B. (2007). *The Charcoal Forest: How Fire Helps Animals and Plants.* Mountain Press Publishing Company, Missoula, MT. See more at <http://www.brucebyersconsulting.com/colorado-fires-and-firemoths/#sthash.qbaxkYbE.dpuf>.

15. Bond, W.J. and Midgley, J.J. (1995). Kill thy neighbour: an individualistic argument for the evolution of flammability. *Oikos* 73, 79–85.

16. See Pyne, S.J. (2001). *Fire: A Brief History.* University of Washington Press, Seattle, WA; and Roos, C.I., Bowman, D.M.J.S., Balch, J.K., Artaxo, P., Bond, W.J., Cochrane, M., D'Antonio, C.M., DeFries, R., Mack, M., Johnston, F.H., Krawchuk, M.A., Kull, C.A., Moritz, M.A., Pyne, S., Scott, A.C., and Swetnam, T.M. (2014). Pyrogeography, historical ecology, and the human dimensions of fire regimes. *Journal of Biogeography* 41, 833–6.

CHAPTER 2

1. <https://www.thenakedscientists.com/articles/interviews/planet-earth-online-friendly-fires>.

2. Hooke, R. (1665). *Micrographia or some physiological descriptions of minute bodies made by magnifying glasses with observations and inquiries thereupon.* Royal Society, London. Observ. XVI. Of Charcoal, or burnt Vegetables.

3. Lyell, C. (1847). On the structure and probable age of the coalfield of the James River, near Richmond, Virginia. *Quarterly Journal of the Geological Society of London* 3, 261–88. See discussion in Scott, A.C. (1998). The legacy of Charles Lyell: advances in our knowledge of coal and coal-bearing strata. In: Blundell, D.J. and Scott, A.C. (eds), *Lyell: The Past is the Key to the Present,* pp. 243–60. Geological Society Special Publication 143.

4. Stopes, M.C. (1919). On the four visible ingredients in banded bituminous coal. *Proceedings of the Royal Society Series B* 90, 470–87.

5. <https://www.mariestopes.org.uk/aboutmariestopesuk>.

6. A discussion on the term fusain is given in Scott, A.C. (1989). Observations on the nature and origin of fusain. *International Journal of Coal Geology* 12, 443–75.

7. Harris, T.M. (1958). Forest fire in the Mesozoic. *Journal of Ecology* 46, 447–53.

8. Harris, T.M. (1981). Burnt ferns from the English Wealden. *Proceedings of the Geologists' Association* 92, 47–58.

9. Hooke (1665). Observ XVII. Of Petrify'd wood, and other Petrify'd bodies. For downloadable text see <https://www.gutenberg.org/files/15491/15491-h/15491-h.htm>.

10. See discussion in Scott (1998).

11. McGinnes, E.A., Harlow, C.A., and Beale, F.C. (1976). Use of scanning electron microscopy and image processing in wood charcoal studies. *Scanning Electron Microscopy* 7, 543–8.

12. Muir, M. (1970). A new approach to the study of fossil wood. *Proceedings of the Third Annual Scanning Electron Microscope Symposium*, ITT Research Institute, Chicago, IL, pp. 129–35.

13. Scott, A. (1974). The earliest conifer. *Nature* 251, 707–8; Scott, A.C. and Chaloner, W.G. (1983). The earliest fossil conifer from the Westphalian B of Yorkshire. *Proceedings of the Royal Society of London B* 220, 163–82.

14. See my picture in Willis, K.J. and McElwain, J.C. (2014). *The Evolution of Plants*, 2nd edition, fig. 3.9, p. 66. Oxford University Press, Oxford.

15. Friis, E.-M. and Skarby, A. (1981). Structurally preserved angiosperm flowers from the Upper Cretaceous of southern Sweden. *Nature* 291, 484–6.

16. Scott, A.C., Cripps, J.A., Nichols, G.J., and Collinson, M.E. (2000). The taphonomy of charcoal following a recent heathland fire and some implications for the interpretation of fossil charcoal deposits. *Palaeogeography, Palaeoclimatology, Palaeoecology* 164, 1–31.

17. Nichols, G.J. and Jones, T.P. (1992). Fusain in Carboniferous shallow marine sediments, Donegal, Ireland: the sedimentological affects of wildfire. *Sedimentology* 39, 487–502.

18. Scott et al. (2000).

19. Scott, A.C., Galtier, J., Gostling, N.J., Smith, S.Y., Collinson, M.E., Stampanoni, M., Marone, F., Donoghue, P.C.J., and Bengtson, S. (2009). Scanning electron microscopy and synchrotron radiation X-ray tomographic microscopy of 330 million year old charcoalified seed fern fertile organs. *Microscopy and Microanalysis* 15, 166–73.

CHAPTER 3

1. Robinson, J.M. (1989). Phanerozoic O2 variation, fire, and terrestrial ecology. *Palaeogeography, Palaeoclimatology, Palaeoecology* 75, 223–40; Robinson, J.M. (1990). Lignin, land plants, and fungi: biological evolution affecting Phanerozoic oxygen balance. *Geology* 15, 607–10; Robinson, J.M. (1991). Phanerozoic atmospheric reconstructions: a terrestrial perspective. *Palaeogeography, Palaeoclimatology, Palaeoecology* 97, 51–62.

2. Falcon-Lang, H.J. (2000). Fire ecology of the Carboniferous tropical zone. *Palaeogeography, Palaeoclimatology, Palaeoecology* 164, 355–71.

3. Krings, M., Kerp, H., Taylor, T.N., and Taylor, E.L. (2003). How Paleozoic vines and lianas got off the ground: on scrambling and climbing Carboniferous–Early Permian Pteridosperms. *The Botanical Review* 69, 204–24.

4. Benton, M.J. (2003). *When Life Nearly Died: The Greatest Mass Extinction Event of All Time*. Thames and Hudson, London.

5. Nichols, D.J. and Johnson, K.R. (2008). *Plants and the K-T Boundary*. Cambridge University Press, Cambridge.

6. Billings Gazette. (1995). *Yellowstone on Fire*, 2nd edition. Billings Gazette, Billings, MT.

7. Harland, W.B. and Hacker, J.L. (1966). 'Fossil' lightning strikes 250 million years ago. *Advancement of Science* 22, 663–71.

8. Berner, R.A., Beerling, D.J., Dudley, R., Robinson, J.M., and Wildman, R.A. (2003). Phanerozoic atmospheric oxygen. *Annual Review of Earth and Planetary Sciences* 31, 105–34.

9. Berner, R.A. and Canfield, D.E. (1989). A new model for atmospheric oxygen over Phanerozoic time. *American Journal of Science* 289, 333–61.

10. Berner, R.A. (2006). A combined model for Phanerozoic atmospheric O2 and CO2. *Geochemica et Cosmochimica Acta* 70, 5653–64;

Berner, R.A. (2009). Phanerozoic atmospheric oxygen: new results using the GEOCARBSULF model. *American Journal of Science* 309, 603–6.

11. Poulsen, C.J., Tabor, C., and White, J.D. (2015). Long-term climate forcing by atmospheric oxygen concentrations. *Science* 348, 1238–41.

12. Watson, A.J., Lovelock, J.E., and Margulis, L. (1978). Methanogenesis, fires and the regulation of atmospheric oxygen. *Biosystems* 10, 293–8; Watson, A.J. and Lovelock, J.E. (2013). The dependence of flame spread and probability of ignition on atmospheric oxygen. In: C.M. Belcher (ed.), *Fire Phenomena and the Earth System: An Interdisciplinary Guide to Fire Science*, pp. 273–87. John Wiley and Sons, Chichester.

13. Wildman, R.A., Hickey, L.J., Dickinson, M.B., Berner, R.A., Robinson, J.M., Dietrich, M., Essenhigh, R.H., and Wildman, C.B. (2004). Burning of forest materials under Late Paleozoic high atmospheric oxygen levels. *Geology* 32, 457–60.

14. Belcher, C.M. and McElwain, J.C. (2008). Limits for combustion in low O2 redefine paleoatmospheric predictions for the Mesozoic. *Science* 321, 1197–200.

15. Belcher, C.M., Yearsley, J.M., Hadden, R.M., McElwain, J.C., and Rein, G. (2010). Baseline intrinsic flammability of Earth's ecosystems estimated from paleoatmospheric oxygen over the past 350 million years. *Proceedings of the National Academy of Sciences* 107, 22448–53.

16. Scott, A.C. and Glasspool, I.J. (2006). The diversification of Paleozoic fire systems and fluctuations in atmospheric oxygen concentration. *Proceedings of the National Academy of Sciences* 103, 10861–5.

17. Glasspool, I.J. and Scott, A.C. (2010). Phanerozoic concentrations of atmospheric oxygen reconstructed from sedimentary charcoal. *Nature Geoscience* 3, 627–30.

CHAPTER 4

1. Glasspool, I.J., Edwards, D., and Axe, L. (2004). Charcoal in the Silurian as evidence of the earliest wildfire. *Geology* 32, 381–3.

2. Glasspool, I.J., Edwards, D., and Axe, L. (2006). Charcoal in the Early Devonian: a wildfire-derived Konservat-Lagerstätte. *Review of Palaeobotany and Palynology* 142, 131–6.

3. Hueber, F.M. (2001). Rotted wood-alga-fungus: the history and life of Prototaxites Dawson 1859. *Review of Palaeobotany and Palynology* 116, 123–58.

4. Scott, A.C. (2010). Charcoal recognition, taphonomy and uses in palaeoenvironmental analysis. *Palaeogeography, Palaeoclimatology, Palaeoecology* 291, 11–39.

5. Scott, A.C. and Glasspool, I.J. (2006). The diversification of Paleozoic fire systems and fluctuations in atmospheric oxygen concentration. *Proceedings of the National Academy of Sciences* 103, 10861–5.

6. Rimmer, S.M., Hawkins, S.J., Scott, A.C., and Cressler, III, W.L. (2015). The rise of fire: fossil charcoal in late Devonian marine shales as an indicator of expanding terrestrial ecosystems, fire, and atmospheric change. *American Journal of Science* 315, 713–33. I was really pleased when our pictures appeared on the front cover of the journal and I was able to dedicate the paper to the memory of Karl Turekian, who had taken such interest in our work on fire through time and who had recently died.

7. Falcon-Lang, H.J. (1998). The impact of wildfire on an Early Carboniferous coastal system, North Mayo, Ireland. *Palaeogeography, Palaeoclimatology, Palaeoecology* 139, 121–38.

8. Rolfe, W.D.I., Durant, G.M., Fallick, A.E., Hall, A.J., Large, D.J., Scott, A.C., Smithson, T.R., and Walkden, G.M. (1990). An early terrestrial biota preserved by Visean vulcanicity in Scotland. In: M.G. Lockley and A. Rice (eds), *Volcanism and Fossil Biotas*. Geological Society of America Special Publication 244, 13–24.

9. Smithson, T.R. (1989). The earliest known reptile. *Nature* 342, 676–8; Smithson, T.R. and Rolfe, W.D.I. (1990). *Westlothiana* gen. nov.: naming the earliest known reptile. *Scottish Journal of Geology* 26, 137–8; Smithson, T.R., Carroll, R.L., Panchen, A.L., and Anderson, S.M. (1994). *Westlothiana lizziae* from the Viséan of East Kirkton, West Lothian, Scotland. *Transactions of the Royal Society of Edinburgh: Earth Sciences* 84, 387–412.

10. Lyell, C. and Dawson, J.W. (1853). On the remains of a reptile (Dendrerpeton acadianum, Wyman and Owen), and of a land shell discovered in the interior of an erect fossil tree in the coal measures of Nova Scotia. *Quarterly Journal of the Geological Society* 9, 58–63.

11. Scott, A.C. (2001). Roasted alive in the Carboniferous. *Geoscientist* 11(3), 4–7.

12. Hudspith, V., Scott, A.C., Collinson, M.E., Pronina, N., and Beeley, T. (2012). Evaluating the extent to which wildfire history can be interpreted from inertinite distribution in coal pillars: an example from the Late Permian, Kuznetsk Basin, Russia. *International Journal of Coal Geology* 89, 13–25.

13. Shao, L., Wang, H., Yu, X., Lu, J., and Mingquan, Z. (2012). Paleo-fires and atmospheric oxygen levels in the latest Permian: evidence from maceral compositions of coals in Eastern Yunnan, Southern China. *Acta Geologica Sinica* (English edition) 86, 949–62.

14. Whiteside, J.H., Lindstrom, S., Irmis, R.B., Glasspool, I.J., Schaller, F., Dunlavey, M., Nesbitt, S.J., Smith, N.D., and Turner, A.H. (2015). Extreme ecosystem instability suppressed tropical dinosaur dominance for 30 million years. *Proceedings of the National Academy of Sciences* 112, 7909–13.

CHAPTER 5

1. Harris, T.M. (1958). Forest fire in the Mesozoic. *Journal of Ecology* 46, 447–53.

2. Berner, R.A., Beerling, D.J., Dudley, R., Robinson, J.M., and Wildman, R.A. (2003). Phanerozoic atmospheric oxygen. *Annual Review of Earth and Planetary Sciences* 31, 105–34.

3. Retallack, G.J., Veevers, J.J., and Morante, R. (1996). Global coal gap between Permian–Triassic extinction and Middle Triassic recovery of peat-forming plants. *Geological Society of America Bulletin* 108, 195–207.

4. Sheldon, N.D. and Retallack, G.J. (2002). Low oxygen levels in earliest Triassic soils. *Geology* 30, 919–22.

5. Uhl, D., Jasper, A., Hamad, A.M.B., and Montenari, M., (2008). Permian and Triassic wildfires and atmospheric oxygen levels. *Proceedings of the 1st WSEAS International Conference on Environmental and Geological Science and Engineering (EG'08)*, Environment and Geoscience Book Series: Energy and Environmental Engineering Series, pp. 179–87;

Whiteside, J.H., Lindstrom, S., Irmis, R.B., Glasspool, I.J., Schaller, M.F., Dunlavey, M., Nesbitt, S.J., Smith, N.D., and Turner, A.H. (2015). Extreme ecosystem instability suppressed tropical dinosaur dominance for 30 million years. *Proceedings of the National Academy of Sciences* 112, 7909–13.

6. Harris, T.M. (1957). A Liasso–Rhaetic flora in South Wales. *Proceedings of the Royal Society of London B* 147, 289–308.

7. Havlik, P., Aiglstorfer, M., El Atfy, H., and Uhl, D. (2013). A peculiar bonebed from the Norian Stubensandstein (Löwenstein Formation, Late Triassic) of southern Germany and its palaeoenvironmental interpretation. *Neues Jahrbuch für Geologie und Paläontologie* 269(3), 321–37.

8. Belcher, C.M., Mander, L., Rein, G., Jervis, F.X., Haworth, M., Hesselbo, S.P., Glasspool, I.J., McElwain, J.C. (2010). Increased fire activity at the Triassic/Jurassic boundary in Greenland due to climate-driven floral change. *Nature Geoscience* 3, 426–9.

9. Petersen, H.I. and Lindström, S. (2012). Synchronous wildfire activity rise and mire deforestation at the Triassic–Jurassic boundary. *PLoS ONE* 7(10), e47236.

10. Berner, R.A. (2009). Phanerozoic atmospheric oxygen: new results using the GEOCARBSULF model. *American Journal of Science* 309, 603–6.

11. Cope, M.J. (1993). A preliminary study of charcoalfield plant fossils from the Middle Jurassic Scalby Formation of North Yorkshire. *Special Papers in Palaeontology* 49, 101–11.

12. Jones, T.P. (1997). Fusain in Late Jurassic sediments from the Witch Ground Graben, North Sea, UK. In: G.F.W. Herngreen (ed.), *Proceedings of the 4th European Palaeobotanical and Palynological Conference: Heerlen/Kerkrade, 19–23 September 1994.* Mededelingen Nederlands Instituut voor Toegepaste Geowetenschappen TNO 58, 93–103.

13. Uhl, D., Jasper, A., and Schweigert, G. (2012). Charcoal in the Late Jurassic (Kimmeridgian) of western and central Europe: palaeoclimatic and palaeoenvironmental significance. *Palaeobiodiversity and Palaeoenvironments* 92, 329–41.

14. Francis, J.E. (1983). The dominant conifer of the Jurassic Purbeck Fm, England. *Palaeontology* 26, 277–94; Francis, J.E. (1984). The seasonal environment of the Purbeck (Upper Jurassic) fossil forests. *Palaeogeography, Palaeoclimatology, Palaeoecology* 48, 285–307.

15. Matthewman, R., Cotton, L.J., Martins, Z., and Sephton, M.A. (2012). Organic geochemistry of late Jurassic paleosols (dirt beds) of Dorset, UK. *Marine and Petroleum Geology* 37, 41–52.

16. Seward, A.C. (1894). The Wealden flora I. Thallophyta-Pteridophyta. *Catalogue of the Mesozoic plants in the Department of Geology, British Museum (Natural History)*, volume 1; Seward, A.C. (1895). The Wealden flora II. Gymnospermae. *Catalogue of the Mesozoic plants in the Department of Geology, British Museum (Natural History)*, volume 2; Seward, A.C. (1913). Contributions to our knowledge of Wealden floras, with especial reference to a collection of plants from Sussex. *Journal of the Geological Society* 69, 85–116; Stopes, M.C. (1915). *Catalogue of the Cretaceous plants in the British Museum (Natural History)*, Part 2; Allen, P. (1976). Wealden of the Weald: a new model. *Proceedings of the Geologists' Association* 86 (for 1975), 389–437; Allen, P. (1981). Pursuit of Wealden models. *Journal of the Geological Society* 138, 375–405.

17. Alvin, K.L. (1974). Leaf anatomy of *Weichselia* based on fusainized material. *Palaeontology* 17, 587–98.

18. Allen (1976).

19. Collinson, M.E., Featherstone, C., Cripps, J.A., Nichols, G.J., and Scott, A.C. (2000). Charcoal-rich plant debris accumulations in the lower Cretaceous of the Isle of Wight, England. *Acta Palaeobotanica* Supplement 2, 93–105.

20. Scott, A.C. and Stea, R. (2002). Fires sweep across the Mid-Cretaceous landscape of Nova Scotia. *Geoscientist* 12(1), 4–6.

21. Falcon-Lang, H.J., Mages, V., and Collinson, M.E. (2016). The oldest *Pinus* and its preservation by fire. *Geology* 44, 303–6.

22. Crane, P.R., Friis, E.M., and Chaloner, W.G. (eds) (2010). Darwin and the evolution of flowers. *Philosophical Transactions of the Royal Society B* 365, 345–543.

23. Stopes, M.C. (1912). Petrifactions of the earliest European angiosperms. *Philosophical Transactions of the Royal Society, B* 203, 75–100.

24. Friis, E.M., Crane, P.R., and Pedersen, K.R. (eds) (2011). *Early Flowers and Angiosperm Evolution*. Cambridge University Press, Cambridge.

25. Hughes, N.F., Drewry, G., and Laing, J.F. (1979). Barremian earliest angiosperm pollen. *Palaeontology* 22, 513–36; Hughes, N.F. and McDougall, A.B. (1990). Barremian-Aptian angiospermid pollen records from southern England. *Review of Palaeobotany and Palynology* 65, 145–51.

26. Hickey, L.J. and Doyle, J.A. (1977). Early Cretaceous fossil evidence for angiosperm evolution. *Botanical Review* 43, 2–104.

27. Friis, E.M., Pedersen, K.R., and Crane, P.R. (2006). Cretaceous angiosperm flowers: innovation and evolution in plant reproduction. *Palaeogeography, Palaeoclimatology, Palaeoecology* 232, 251–93; Friis, E.M., Pedersen, K.R., and Crane, P.R. (2010). Diversity in obscurity: fossil flowers and the early history of angiosperms. *Philosophical Transactions of the Royal Society B* 365, 369–82; Brown, S.A.E., Scott, A.C., Glasspool, I.J., and Collinson, M.E. (2012). Cretaceous wildfires and their impact on the Earth system. *Cretaceous Research* 36, 162–90.

28. Eklund, H. (2003). First Cretaceous flowers from Antarctica. *Review of Palaeobotany and Palynology* 127, 187–217; Eklund, H., Cantrill, D.J., and Francis, J.E. (2004). Late Cretaceous plant mesofossils from Table Nunatak, Antarctica. *Cretaceous Research* 25, 211–28.

29. Bond, W.J. and Scott, A.C. (2010). Fire and the spread of flowering plants in the Cretaceous. *New Phytologist* 118, 1137–50. See our press release: <https://www.royalholloway.ac.uk/research/news/newsarticles/firefuelsflowerssuccess.aspx>.

30. Evans, D.C., Eberth, D.A., and Ryan, M.J. (2015). Hadrosaurid (*Edmontosaurus*) bonebeds from the Horseshoe Canyon Formation (Horsethief Member) at Drumheller, Alberta, Canada: geology, preliminary taphonomy, and significance. *Canadian Journal of Earth Sciences* 52, 642–54.

31. Keeley, J.E., Pausas, J.G., Rundel, P.W., Bond, W.J., and Bradstock, R.A. (2011). Fire as an evolutionary pressure shaping plant traits. *Trends in Plant Science* 16, 406–11.

32. He, T., Pausas, J.G., Belcher, C.M., Schwilk, D.W., and Lamont, B.B. (2012). Fire-adapted traits of *Pinus* arose in the fiery Cretaceous. *New Phytologist* 194, 751–9.

33. He, T., Lamont, B.B., and Downes, K.S. (2011). *Banksia* born to burn. *New Phytologist* 191, 184–96; Lamont, B.B. and He, T. (2012). Fire adapted Gondwanan Angiosperm floras evolved in the Cretaceous. *BMC Evolutionary Biology* 12, article 223.

34. Carpenter, R.J., Macphail, M.K., Jordan, G.J., and Hill, R.S. (2015). Fossil evidence for open, Proteaceae-dominated heathlands and fire in the Late Cretaceous of Australia. *American Journal of Botany* 102, 1–16.

35. Kump, L. (1988). Terrestrial feedback in atmospheric oxygen regulation by fire and phosphorous. *Nature* 335, 152–4.

36. Alvarez, L.W., Alvarez, W., Asaro, F., and Michal, H.V. (1980). Extraterrestrial cause for the Cretaceous–Tertiary extinction. *Science* 208, 1095–108; Hildebrand, A.R. et al. (1991). Chicxulub crater: a possible Cretaceous–Tertiary boundary impact crater on the Yucatan Peninsula, Mexico. *Geology* 19, 867–71.

37. Wolbach, W.S., Lewis, R.S., and Anders, E. (1985). Cretaceous extinctions: evidence for wildfires and search for meteoritic material. *Science* 230, 167–230; Wolbach, W.S., Gilmour, I., Anders, E., Orth, C.J., and Brooks, R.R. (1988). Global fire at the Cretaceous–Tertiary boundary. *Nature* 334, 665–9; Wolbach, W.S., Gilmour I., and Anders, E. (1990). Major wildfires at the Cretaceous/Tertiary. *Geological Society of America Special Paper* 247, 391–400.

38. Jones, T.P. and Lim, B. (2000). Extraterrestrial impacts and wildfires. *Palaeogeography, Palaeoclimatology, Palaeoecology* 164, 57–66.

39. Scott, A.C., Lomax, B.H., Collinson, M.E., Upchurch, G.R., and Beerling, D.J. (2000). Fire across the K/T boundary: initial results from the Sugarite Coal, New Mexico, USA. *Palaeogeography, Palaeoclimatology, Palaeoecology* 164, 381–95.

40. Hildebrand et al. (1991).

41. Belcher, C.M., Collinson, M.E., Sweet, A.R., Hildebrand, A.R., and Scott, A.C. (2003). Fireball passes and nothing burns. The role of thermal radiation in the Cretaceous–Tertiary event: evidence from the charcoal record of North America. *Geology* 31, 1061–4.

42. Belcher, C.M., Collinson, M.E., and Scott, A.C. (2005). Constraints on the thermal energy released from the Chicxulub impactor: new evidence from multi-method charcoal analysis. *Journal of the Geological Society* 162, 591–602.

43. Melosh, H.J., Schneider, N.M., Zahnle, K.J., and Latham, D. (1990). Ignition of global wildfires at the Cretaceous/Tertiary boundary. *Nature* 343, 251–4; Belcher, C.M. (2009). Reigniting the Cretaceous–Palaeogene firestorm debate. *Geology* 37, 1147–8.

44. Harvey, M.C., Brassell, S.C., Belcher, C.M., and Montanari, A. (2008). Combustion of fossil organic matter at the K–P boundary. *Geology* 36, 335–58.

45. Belcher, C.M., Finch, P., Collinson, M.E., Scott, A.C., and Grassineau, N.V. (2009). Geochemical evidence for combustion of hydrocarbons during the K–T impact event. *Proceedings of the National Academy of Sciences* 106, 4112–17.

46. As I finish writing this, again there is a new claim of global fire. Scientists will continue to argue. Toon, O.B., Bardeen, C., and Garcia, R. (2016). Designing global climate and atmospheric chemistry simulations for 1 and 10km diameter asteroid impacts using the properties of ejecta from the K–Pg impact. *Atmospheric Chemistry And Physics* 16, 13185–212.

CHAPTER 6

1. Kennett, J.P. and Stott, L.D. (1991). Abrupt deep-sea warming, palaeoceanographic changes and benthic extinctions at the end of the Palaeocene. *Nature* 353, 225–9.

2. Dickens, G.R. (2003). Rethinking the global carbon cycle with a large, dynamic and microbially mediated gas hydrate capacitor. *Earth and Planetary Science Letters* 213, 169–83; Kurtz, A.C., Kump, L.R., Arthur, M.A., Zachos, J.C., and Paytan, A. (2003). Early Cenozoic decoupling of the global carbon and sulfur cycles. *Paleoceanography* 18, article 1090; Sluijs, A., Schouten, S., Pagani, M., Woltering, M., Brinkhuis, H., Damsté, J.S.S., Dickens, G.R., Huber, M., Reichart, G.-J., and Stein, R. (2006). Subtropical Arctic Ocean temperatures during the Palaeocene/Eocene Thermal Maximum. *Nature* 441, 610–13.

3. Finkelstein, D.B., Pratt, L.M., and Brassell, S.C. (2006). Can biomass burning produce a globally significant carbon-isotope excursion in the sedimentary record? *Earth and Planetary Science Letters* 250, 501–10.

4. Kurtz et al. (2003).

5. Collinson, M.E., Hooker, J.J., and Gröcke, D.R. (2003). Cobham lignite bed and penecontemporaneous macrofloras of southern England: a record of vegetation and fire across the Paleocene–Eocene Thermal Maximum. In: S.L. Wing, P.D. Gingerich, B. Schmitz, and E. Thomas (eds), *Causes and Consequences of Globally Warm Climates in the Early Paleogene*. Geological Society of America, Special Papers 369, 333–49.

6. Steart, D.C., Collinson, M.E., Scott, A.C., Glasspool, I.J., and Hooker, J.J. (2007). The Cobham lignite bed: the palaeobotany of two petrographically contrasting lignites from either side of the Paleocene–Eocene carbon isotope excursion. *Acta Palaeobotanica* 47, 109–25.

7. See Collinson, M.E., Steart, D.C., Scott, A.C., Glasspool, I.J., and Hooker, J.J. (2007). Episodic fire, runoff and deposition at the Palaeocene–Eocene boundary. *Journal of the Geological Society* 164, 87–97.

8. Steart et al. (2007).

9. Bowen, G.J., Beerling, D.J., Koch, P.L., Zachos, J.C., and Quattlebaum, T.A. (2004). Humid climate state during the Palaeocene/Eocene Thermal Maximum. *Nature* 432, 495–9; Schmitz, B. and Pujalte, V. (2007). Abrupt increase in seasonal extreme precipitation at the Paleocene–Eocene boundary. *Geology* 35, 215–18.

10. Collinson, M.E., Steart, D.C., Harrington, G.J., Hooker, J.J., Scott, A.C., Allen, L.O., Glasspool, I.J., and Gibbons, S.J. (2009). Palynological evidence of vegetation dynamics in response to palaeoenvironmental change across the onset of the Paleocene–Eocene Thermal Maximum at Cobham, Southern England. *Grana* 48, 38–66.

11. Collinson et al. (2009).

12. Pancost, R.D., Steart, D.S., Handley, L., Collinson, M.E., Hooker, J.J., Scott, A.C., Grassineau, N.J., and Glasspool, I.J. (2007). Increased terrestrial methane cycling at the Palaeocene–Eocene Thermal Maximum. *Nature* 449, 332–5.

13. Riegel, W., Wilde, V., and Lenz, O.K. (2012). The early Eocene of Schöningen (N-Germany): an interim report. *Austrian Journal of Earth Sciences* 105, 88–109; Robson, B.E., Collinson, M.E., Riegel, W., Wilde, V., Scott, A.C., and Pancost, R.D. (2014). A record of fire through the Early Eocene. *Rendiconti Online della Società Geologica Italiana* 31, 187–8.

14. Inglis, G.N., Collinson, M.E., Riegel, W., Wilde, V., Farnsworth, A., Lunt, D.J., Valdes, P., Robson, B.E., Scott, A.C., Lenz, O.K., Naafs, D.A., and Pancost, R.D. (2017). Mid-latitude continental temperatures through the early Eocene in Western Europe. *Earth and Planetary Science Letters* 460, 86–96.

15. Robson et al. (2014).

16. Holdgate, G.R., Wallace, M.W., Sluiter, I.R.K., Marcuccioa, D., Fromhold, T.A., and Wagstaff, B.E. (2014). Was the Oligocene–Miocene a time of fire and rain? Insights from brown coals of the

southeastern Australia Gippsland Basin. *Palaeogeography, Palaeo-climatology, Palaeoecology* 411, 65–78.

17. Herring, J.R. (1985). Charcoal fluxes into sediments of the North Pacific Ocean: the Cenozoic record of burning. In: *The Carbon Cycle and Atmospheric CO2: Natural Variations Archean to Present*. Geophysical Monographs 32, 419–42.

18. Cerling, T.E., Wang, Y., and Quade, J. (1993). Expansion of C4 ecosystems as an indicator of global ecological change in the late Miocene. *Nature* 361, 344–5.

19. Urban, M.A., Nelson, D.M., Street-Perrott, F.A., Verschuren, D., and Hu, F.S. (2015). A late-Quaternary perspective on atmospheric pCO2, climate, and fire as drivers of C4-grass abundance. *Ecology* 96, 642–53.

20. Bond, W.J., Woodward, F.I., and Midgley, G.F. (2005). The global distribution of ecosystems in a world without fire. *New Phytologist* 165, 525–38; Keeley, J.E. and Rundel, P.W. (2005). Fire and the Miocene expansion of C4 grasslands. *Ecology Letters* 8, 683–90; Osborne, C.P. (2008). Atmosphere, ecology and evolution: what drove the Miocene expansion of C-4 grasslands? *Journal of Ecology* 96, 35–45; Beerling, D.J. and Osborne, C.P. (2006). Origin of the savanna biome. *Global Change Biology* 12, 2023–31; Staver, A.C., Archibald, S., and Levin, S.A. (2011). The global extent and determinants of savanna and forest as alternative biome states. *Science* 334, 230–2.

21. Thorn, V.C. and DeConto, R. (2006). Antarctic climate at the Eocene/Oligocene boundary: climate model sensitivity to high latitude vegetation type and comparisons with the palaeobotanical record. *Palaeogeography, Palaeoclimatology, Palaeoecology* 231, 134–57; Francis, J.E. and Hill, R.S. (1996). Fossil plants from the Pliocene Sirius Group, transantarctic mountains: evidence for climate from growth rings and fossil leaves. *PALAIOS* 11, 389–96.

22. Hill, D.J., Haywood, A.M., Valdes, P.J., Francis, J.E., Lunt, D.J., Wade, B.S., and Bowman, V.C. (2013). Paleogeographic controls on the onset of the Antarctic circumpolar current. *Geophysical Research Letters* 40, 5199–204; Siegert, M.J., Barrett, P., Decont, R., Dunbar, R., Cofaigh, C.O., Passchier, S., and Naish, T. (2008). Recent advances in understanding Antarctic climate evolution. *Antarctic Science* 20, 313–25.

23. <http://www.gpwg.org/gpwgdb.html>.

24. Swetnam, T.W. (1993). Fire history and climate change in giant sequoia groves. *Science* 262, 885–9.

25. Westerling, A.L., Hidalgo, H.G., Cayan, D.R., and Swetnam, T.W. (2006). Warming and earlier spring increase western U.S. forest wildfire activity. *Science*, 313, 940–3.

26. Marlon, J.R., Bartlein, P.J., Walsh, M.K., Harrison, S.P., Brown, K.J., Edwards, M.E., Higuera, P.E., Power, M.J., Anderson, R.S., Briles, C., Brunelle, A., Carcaillet, C., Daniels, M., Hu, F.S., Lavoie, M., Long, C., Minckley, T., Richard, P.J.H., Scott, A.C., Shafer, D.S., Tinner, W., Umbanhowar, C.E., Jr, and Whitlock, C. (2009). Wildfire responses to abrupt climate change in North America. *Proceedings of the National Academy of Sciences* 106, 2519–24.

27. Kerr, R.A. (2007). Mammoth-killer impact gets mixed reception from Earth scientists. *Science* 316, 1264–5; Firestone, R.B., West, A., Kennett, J.P., et al. (2007). Evidence for an extraterrestrial impact 12,900 years ago that contributed to the megafaunal extinctions and the Younger Dryas cooling. *Proceedings of the National Academy of Sciences* 104, 16016–21. The Younger Dryas was the last of three closely related cooling events that took place over the past 16,000 years, following the end of the last ice age, 27,000–24,000 years ago. It is named after a flower that became common in Europe during this time (*Dryas octopetala*), which thrives in cold conditions.

28. Firestone et al. (2007).

29. Kennett, D.J., Kennett, J.P., West, C.J., et al. (2008). Wildfire and abrupt ecosystem disruption on California's Northern Channel Islands at the Allerod–Younger Dryas boundary (13.0–12.9 ka). *Quaternary Science Reviews* 27, 2530–45.

30. Pinter, N., Scott, A.C., Daulton, T.L., Podoll, A., Koeberl, C., Anderson, R.S., and Ishman, S.E. (2011). The Younger Dryas impact hypothesis: a requiem. *Earth Science Reviews* 106, 247–64.

31. Kennett, D.J., Kennett, J.P., West, A., et al. (2009). Nanodiamonds in the Younger Dryas boundary sediment layer. *Science* 323, 94. But see Daulton, T.L., Amari, S., Scott, A.C., Hardiman, M., Pinter, N., and Anderson, R.S. (2017). Comprehensive analysis of nanodiamond evidence relating to the Younger Dryas impact hypothesis. *Journal of Quaternary Science* 32, 7–34.

32. Scott, A.C., Pinter, N., Collinson, M.E., Hardiman, M., Anderson, R.S., Brain, A.P.R., Smith, S.Y., Marone, F., and Stampanoni, M. (2010). Fungus, not comet or catastrophe, accounts for carbonaceous spherules in the Younger Dryas 'impact layer'. *Geophysical Research Letters* 37, L14302.

33. Scott et al. (2010). See also Scott, A.C., Hardiman, M., Pinter, N.P., Anderson, R.S., Daulton, T.L., Ejarque, A., Finch, P., and Carter-Champion, A. (2017). Interpreting palaeofire evidence from fluvial sediments: a case study from Santa Rosa Island, California with implications for the Younger Dryas impact hypothesis. *Journal of Quaternary Science* 32, 35–47.

34. Daulton, T.L., Pinter, N., and Scott, A.C. (2010). No evidence of nanodiamonds in Younger Dryas sediments to support an impact event. *Proceedings of the National Academy of Sciences* 107, 16043–7. See also Daulton et al. (2017).

CHAPTER 7

1. Gowlett, J. (2010). Firing up the social brain. In: R. Dunbar, C. Gamble, and J. Gowlett (eds), *Social Brain and Distributed Mind*, pp. 345–70. The British Academy, London; Gowlett, J. and Wrangham, R.W. (2013). Earliest fire in Africa: the convergence of archaeological evidence and the cooking hypothesis. *Azania: Archaeological Research in Africa* 48, 5–30; Twomey, T. (2013). The cognitive implications of controlled fire use by early humans. *Cambridge Archaeological Journal* 23, 113–28; Dunbar, R.I.M. and Gowlett, J. (2014). Fireside chat: the impact of fire on hominin socioecology. In: R.I.M. Dunbar, C. Gamble, and J. Gowlett (eds), *Lucy to Language: The Benchmark Papers*, pp. 277–96. Oxford University Press, Oxford; Smith, A.R., Carmody, R.N., Dutton, R.J., et al. (2015). The significance of cooking for early hominin scavenging. *Journal of Human Evolution* 84, 62–70.

2. Gowlett and Wrangham (2013); Gowlett, J. (2010). Firing up the social brain. In: R. Dunbar, C. Gamble, and J. Gowlett (eds), *Social Brain and Distributed Mind*, pp. 345–70. Oxford University Press, Oxford, table 17.1, p. 349.

3. Warneken, F. and Rosati, A.G. (2015). Cognitive capacities for cooking in chimpanzees. *Proceedings of the Royal Society B* 282, 1809.

4. Wrangham, R. (2009). *Catching Fire: How Cooking Made Us Human.* Basic Books, New York; Rowlett, R.M. (2000). Fire control by *Homo erectus* in East Africa and Asia. *Acta Anthropologica Sinica,* Supplement to 19, 198–208; Clark, J.D and Harris, J.W.K. (1985). Fire and its roles in early hominid lifeways. *African Archaeological Review* 3, 3–27.

5. Zhong, M., Shi, C., Gao, X., Wu, X., Chen, F., Zhang, S., Zhang, X., and Olsen, J.W. (2014). On the possible use of fire by *Homo erectus* at Zhoukoudian, China. *Chinese Science Bulletin* 59(3), 335–43.

6. Roebroeks, W. and Villa, P. (2011). On the earliest evidence for habitual use of fire in Europe. *Proceedings of the National Academy of Sciences* 108, 5209–14.

7. Bellomo, R.V. (1993). A methodological approach for identifying archaeological evidence of fire resulting from human activities. *Journal of Archaeological Science* 20, 525–55.

8. Dibble, H., Berna, F., Goldberg, P., McPherron, S.J.P., Mentzer, S., Niven, L., et al. (2009). A preliminary report on Pech de l'Azé IV, Layer 8 (Middle Paleolithic, France). *PaleoAnthropology,* 182–219, see p. 187.

9. Mentzer, S.M. (2012). Microarchaeological approaches to the identification and interpretation of combustion features in prehistoric archaeological sites. *Journal of Archaeological Method and Theory* 21, 616–68.

10. Berna, F., Goldberg, P., Horwitz, L.K., Brink, J., Holt, S., Bamford, M., and Chazang, M. (2001). Microstratigraphic evidence of *in situ* fire in the Acheulean strata of Wonderwerk Cave, Northern Cape province, South Africa. *Proceedings of the National Academy of Sciences* 109(20), E1215–E1220.

11. James, S.R. (1989). Hominid use of fire in the Lower and Middle Pleistocene: a review of the evidence. *Current Anthropology* 30, 1–26; Sandgathe, D.M., Dibble, H.L., Goldberg, P., McPherron, S.P., Turq, A., Niven, L., and Hodgkins, J. (2011). Timing of the appearance of habitual fire use. *Proceedings of the National Academy of Sciences* 108, E298.

12. Roebroeks and Villa (2011).

13. See also, Goren-Inbar, N., Alperson, N., Kislev, M.E., Simchoni, O., Melamed, Y., Ben-Nun, A., and Werker, E. (2004). Evidence of hominin control of fire at Gesher Benot Ya'aqov, Israel. *Science* 304, 725–7; Alperson-Afil, N. (2008). Continual fire-making by hominins at Gesher Benot Ya'aqov, Israel. *Quaternary Science Reviews* 27, 1733–9.

14. Shimelmitz, R., Kuhn, S.L., Jelinek, A.J., et al. (2014). 'Fire at will': the emergence of habitual fire use 350,000 years ago. *Journal of Human Evolution* 77, 196–203.

15. Shahack-Gross, R., Berna, F., Karkanas, P., Lemorini, C., Gopher, A., and Barkai, R. (2014). Evidence for the repeated use of a central hearth at Middle Pleistocene (300 ky ago) Qesem Cave, Israel. *Journal of Archaeological Science* 44, 12–21.

16. Thieme, H. (1998). The oldest spears in the world: Lower Palaeolithic hunting weapons from Schöningen, Germany. In: E. Carbonell, J.M. Bermudez de Castro, J.L. Arsuaga, and X.P. Rodriguez (eds), *The First Europeans: Recent Discoveries and Current Debate*, pp. 169–93. Aldecoa, Burgos; Stahlschmidt, M.C., Miller, C.E., Ligouis, B., Hambach, U., Goldberg, P., Berna, F., Richter, D., Urban, B., Serangeli, J., and Conard, N.J. (2015). On the evidence for human use and control of fire at Schöningen. *Journal of Human Evolution* 89, 181–201.

17. See description by Preece, R.C., Gowlett, J., Parfitt, S.A., Bridgland, D.R., and Lewis, S.G. (2006). Humans in the Hoxnian: habitat, context and fire use at Beeches Pit, West Stow, Suffolk, UK. *Journal Of Quaternary Science* 21(5), 485–96.

18. See Gowlett and Wrangham (2013) for recent discussion.

19. Bensten, S.E. (2014). Using pyrotechnology: fire-related features and activities with a focus on the African Middle Stone Age. *Journal of Archaeological Research* 22, 141–75.

20. Karkanas, P., et al. (2007). Evidence for habitual use of fire at the end of the Lower Paleolithic: site-formation processes at Qesem Cave, Israel. *Journal of Human Evolution* 53, 197–212; Alperson-Afil, N. and Goren-Inbar, N. (2010). *The Acheulian Site of Gesher Benot Ya'aqov: Ancient Flames and Controlled Use of Fire*. Springer, New York, volume 2; Roos, C.I., Bowman, D.M.J.S., Balch, J.K., Artaxo, P., Bond, W.J., Cochrane, M., D'Antonio, C.M., DeFries, R., Mack, M., Johnston, F.H., Krawchuk, M.A., Kull, C.A., Moritz, M.A., Pyne, S., Scott, A.C., and Swetnam, T.M. (2014). Pyrogeography, historical ecology, and the human dimensions of fire regimes. *Journal of Biogeography* 41, 833–6.

21. Sorensen, A., Roebroeks, W., and van Gijn, A. (2014). Fire production in the deep past? The expedient strike-a-light model. *Journal of Archaeological Science* 42, 476–86, p. 477.

22. Comment by Chris Stringer on <http://www.bbc.co.uk/news/science-environment-32976352>.

23. Koller, J., Baumer, U., and Mania, D. (2001). High-tech in the Middle Palaeolithic: Neandertal-manufactured pitch identified. *European Journal of Archaeology* 4, 385–97.

24. Jones, M. (2002). *The Molecule Hunt: Archaeology and the Search for Ancient DNA.* Allen Lane, London.

25. Brown, T., Allaby, R., Sallares, R., and Jones, G. (1998). Ancient DNA in charred wheats: taxonomic identification of mixed and single grains. *Ancient Biomolecules* 2, 185–93.

26. Brown, T.A., Cappellini, E., Kistler, L., Lister, D.L., Oliveira, H.R., Wales, N., and Sclumbaum, A. (2015). Recent advances in ancient DNA research and their implications for archaeobotany. *Vegetation History and Archaeobotany* 24, 207–14; Fernandez, E., Thaw, S., Brown, T.A., Arroyo-Pardo, E., Buxó, R., Serret, M.D., and Araus, J.L. (2013). DNA analysis in charred grains of naked wheat from several archaeological sites in Spain. *Journal of Archaeological Science* 40, 659–70; Brown, T.A. and Brown, K.A. (2011). *Biomolecular Archaeology: An Introduction.* Wiley-Blackwell, Chichester; Brown, T.A. (1999). How ancient DNA may help in understanding the origin and spread of agriculture. *Philosophical Transactions of the Royal Society B* 354, 89–98.

27. Brown et al. (2015).

28. Margaritis, E. and Jones, M. (2006). Beyond cereals: crop processing and Vitis vinifera L. Ethnography, experiment and charred grape remains from Hellenistic Greece. *Journal of Archaeological Science* 33(6), 784–805.

29. Chrzazvez, J., Thery-Parisot, I., Fiorucci, G., Terral, J.F., and Thibaut, B. (2014). Impact of post-depositional processes on charcoal fragmentation and archaeobotanical implications: experimental approach combining charcoal analysis and biomechanics. *Journal of Archaeological Science* 44, 30–42; Henry, A. and Thery-Parisot, I. (2014). From Evenk campfires to prehistoric hearths: charcoal analysis as a tool for identifying the use of rotten wood as fuel. *Journal of Archaeological Science* 52, 321–36; Thery-Parisot, I. and Henry, A. (2012). Seasoned or green? Radial cracks analysis as a method for identifying the use of green wood as fuel in archaeological charcoal. *Journal of Archaeological Science* 39, 381–8.

30. Bliege Bird, R., Bird, D.W., Codding, B.F., Parker, C.H., and Jones, J.H. (2008). The 'fire stick farming' hypothesis: Australian Aboriginal foraging strategies, biodiversity, and anthropogenic fire mosaics. *Proceedings of the National Academy of Sciences* 105, 14796–801.

31. Archibald, S., Staver, A.C., and Levin, S.A. (2012). Evolution of human driven fire regimes in Africa. *Proceedings of the National Academy of Sciences* 109, 847–52; Pyne, S.J. (1992). *Burning Bush. A Fire History of Australia*. Allen and Unwin, Sydney; Pyne, S.J. (2001). *Fire: A Brief History*. University of Washington Press, Seattle.

32. Scott et al. (2014); Archibald, S. (2016). Managing the human component of fire regimes: lessons from Africa. *Philosophical Transactions of the Royal Society B* 371, 20150346.

33. Pyne (1992); Gould, R.A. (1971). Uses and effects of fire among the western desert Aborigines of Australia. *Mankind* 8, 14–24; Williams, A.N., Mooney, S.D., Sisson, S.A., and Marlon, J. (2015). Exploring the relationship between Aboriginal population indices and fire in Australia over the last 20,000 years. *Palaeogeography, Palaeoclimatology, Palaeoecology* 432, 49–57.

34. Swetnam, T.W., Farella, J., Roos, C.I., Liebmann, M.J., Falk, D.A., and Allen, C.D. (2016). Multi-scale perspectives of fire, climate and humans in western North America and the Jemez Mountains, USA. *Philosophical Transactions of the Royal Society B* 371, 20150168.

35. Turney, C.S.M., Kershaw, A.P., Moss, P., et al. (2001). Redating the onset of burning at Lynch's Crater (North Queensland): implications for human settlement in Australia. *Journal of Quaternary Science* 16, 767–71.

36. Daniau, A.-L., d'Errico, F., and Sánchez Goñi, M.F. (2010). Testing the hypothesis of fire use for ecosystem management by Neanderthal and Upper Palaeolithic modern human populations. *PLoS ONE* 5(2), e9157.

37. Hardiman, M., Scott, A.C., Pinter, N.P., Anderson, R.S., Ejarque, A., and Carter-Champion, A. (2016). Fire history on California Channel Islands spanning human arrival in the Americas. *Philosophical Transactions of the Royal Society B* 371, 20150167.

38. Hardiman et al. (2016); Muhs, D.R., Simmons, K.R., Groves, L.T., et al. (2015). Late Quaternary sea-level history and the antiquity of mammoths (*Mammuthus exilis* and *Mammuthus columbi*), Channel Islands National Park, California, USA. *Quaternary Research* 83, 502–21.

39. Balch, J., Nagy, R., Archibald, S., Bowman, D., Moritz, M., Roos, C., Scott, A.C., and Williamson, G. (2016). Global combustion: the connection between fossil fuel and biomass burning emissions (1997–2010). *Philosophical Transactions of the Royal Society B* 371, 20150177.

40. Westerling, A.L., Hidalgo, H.G., Cayan, D.R., and Swetnam, T.W. (2006). Warming and earlier spring increase western U.S. forest wildfire activity. *Science* 313, 940–3; Westerling, A.L., Turner, M.G., Smithwick, E.A.H., Romme, W.H., and Ryan, M.G. (2011). Continued warming could transform Greater Yellowstone fire regimes by mid-21st century. *Proceedings of the National Academy of Sciences* 108, 13165–70; Westerling, A.L.R. (2016). Increasing western US forest wildfire activity: sensitivity to changes in the timing of spring. *Philosophical Transactions of the Royal Society B* 371, 20150178.

CHAPTER 8

1. This was the topic for a recent Royal Society meeting called Fire and Mankind, held in September 2015: Scott, A.C., Chaloner, W.G., Belcher, C., and Roos, C. (eds) (2016). The interaction of fire and mankind. *Philosophical Transactions of the Royal Society B* 371.

2. Collinson, M.E., and Crane, P. R. (1978). *Rhododendron* seeds from Palaeocene of southern England. *Botanical Journal of the Linnean Society* 76(3), 195–205.

3. Pearce, F. (2015). *The New Wild: Why Invasive Species Will Be Nature's Solution.* Icon Books, London.

4. Crisp, M.D., Burrows, G.E., Cook, L.G., Thornhill, A.H., and Bowman, D.M.J.S. (2011). Flammable biomes dominated by eucalypts originated at the Cretaceous–Palaeogene boundary. *Nature Communications* 2, 193.

5. Balch, J.K., Bradley, B.A., D'Antonio, C.M., and Gomez-Dans, J. (2013). Introduced annual grass increases regional fire activity across the arid western USA (1980–2009). *Global Change Biology* 19, 173–83; Butler, D.W., Fensham, R.J., Murphy, B.P., Haberle, S.G., Bury, S.J., and Bowman, D.M.J.S. (2014). Aborigines managed forest, savanna and grassland: biome switching in montane eastern Australia. *Journal of Biogeography* 41, 1492–505.

6. See Scott et al. (2014); Olsson, A.D., Betancourt, J., McClaran, M.P., et al. (2012). Sonoran Desert ecosystem transformation by a C4 grass without the grass/fire cycle. *Diversity and Distributions* 18, 10–21. Springer, A.C., Swann, D.E., and Crimmins, M.A. (2015). Climate change impacts on high elevation saguaro range expansion. *Journal of Arid Environments* 116, 57–62; Brooks, M.L., D'Antonio, C.M., Richardson, D.M., et al. (2004). Effects of invasive alien plants on fire regimes. *Bioscience* 54, 677–88.

7. Mistry, J., Bilbao, B., and Berardi, A. (2016). Engineering and innovation community owned solutions for fire management in tropical forest and savanna ecosystems: case studies from indigenous communities of South America. *Philosophical Transactions of the Royal Society B*, 371, 20150174.

8. Cochrane, M.A. (2003). Fire science for rainforests. *Nature* 421, 913–19; Cochrane, M.A. (ed.) (2009). *Tropical Fire Ecology: Climate Change, Land Use and Ecosystem Dynamics*. Springer, Berlin; Davidson, E.A., de Araujo, A.C., Artaxo, P., et al. (2012). The Amazon basin in transition. *Nature* 481, 321–8; Balch, J.K., Brando, P.M., Nepstad, D.C., et al. (2015). The susceptibility of southeastern Amazon forests to fire: insights from a large-scale burn experiment. *Bioscience* 65, 893–905.

9. Mistry et al. (2016).

10. Nawrotzki, R.J., Brenkert-Smith, H., Hunter, L.M., et al. (2014). Wildfire-migration dynamics: lessons from Colorado's Four Mile Canyon Fire. *Society & Natural Resources* 27, 215–25.

11. Moody, J.A. and Ebel, B.A. (2012). Hyper-dry conditions provide new insights into the cause of extreme floods after wildfire. *Catena* 93, 58–63.

12. The problem of the death of firefighters was highlighted by those who went to extinguish a fire in an area near Phoenix, Arizona in 2013 in the Yarnell Hill Fire. Nineteen firefighters were killed. This was a lightning-started wildfire. <https://en.wikipedia.org/wiki/Yarnell_Hill_Fire>.

13. See a useful discussion in Doerr, S. and Santín, C. (2016). The 'wildfire problem': perceptions and realities in a changing world. *Philosophical Transactions of the Royal Society B* 371, 20150345. This issue has also been highlighted in the highly respected journal *Science*: Topik, C. (2015). Wildfires burn science capacity. *Science* 349, 1263; North, M.P., Stephens, S.L., Collins, B.M., Agee, J.K., Aplet, G., Franklin,

J.F., and Fulé, P.Z. (2015). Reform forest fire management: agency incentives undermine policy effectiveness. *Science* 349, 1280–1.

14. Bowman, D.M.J.S., Balch, J., Artaxo, P., Bond, W.J., Cochrane, M.A., D'Antonio, C.M., DeFries, R., Johnston, F.H., Keeley, J.E., Krawchuk, M.A., Kull, C.A., Mack, M., Moritz, M.A., Pyne, S.J., Roos, C.I., Scott, A.C., Sodhi, N.S., and Swetnam, T.W. (2011). The human dimension of fire regimes on Earth. *Journal of Biogeography* 38, 2223–36; Roos, C.I., Bowman, D.M.J.S., Balch, J.K., Artaxo, P., Bond, W.J., Cochrane, M., D'Antonio, C.M., DeFries, R., Mack, M., Johnston, F.H., Krawchuk, M.A., Kull, C.A., Moritz, M.A., Pyne, S., Scott, A.C., and Swetnam, T.M. (2014). Pyrogeography, historical ecology, and the human dimensions of fire regimes. *Journal of Biogeography* 41, 833–6.

15. Doerr and Santin (2016).

16. Earles, T.A., Wright, K.R., Brown, C., et al. (2004). Los Alamos forest fire impact modeling. *Journal of the American Water Resources Association* 40, 371–84; Holloway, M. (2000). Uncontrolled: the Los Alamos blaze exposes the missing science of forest management. *Scientific American* 283, 16–17.

17. Scott et al. (2014).

18. Johnson, B. (1984). *The great fire of Borneo: report of a visit to Kalimantan-Timur a year later, May 1984*. World Wildlife Fund, Godalming.

19. Bowman, D.M.J.S., et al. (2009). Fire in the Earth System. *Science* 324, 481–4; Scott et al. (2014); Bowman, D.M.J.S., Perry, G., Higgins, S., Johnson, C., and Murphy, B. (2016). Pyrodiversity and biodiversity are coupled because fire is embedded in food-webs. *Philosophical Transactions of the Royal Society B* 371, 20150169; Pringle, R.M., Kimuyu, D.M., Sensenig, R.L., et al. (2015). Synergistic effects of fire and elephants on arboreal animals in an African savanna. *Journal of Animal Ecology* 84, 1637–45; Strahan, R.T., Stoddard, M.T., Springer, J.D., et al. (2015). Increasing weight of evidence that thinning and burning treatments help restore understory plant communities in ponderosa pine forests. *Forest Ecology and Management* 353, 208–20; Keane, R.E., McKenzie, D., Falk, D.A., et al. (2015). Representing climate, disturbance, and vegetation interactions in landscape models. *Ecological Modelling* 309, 33–47.

20. See Keeley, J.E., Bond, W.J., Bradstock, R.A., Pausas, J.G., and Rundel, P.W. (2012). *Fire in Mediterranean Climate Ecosystems: Ecology, Evolution*

and Management. Cambridge University Press, Cambridge; Sugihara, N.G., Van Wagtendonk, J.W., Shaffer, K.E., Fites, K.J., and Thode, A.E. (eds) (2006). *Fire in California's Ecosystems.* University of California Press, Berkeley. There is also a very good website that discusses many of these important issues in a very balanced way: <http://www.californiachaparral.com>; Mortiz, M.A., Batlori, E., Bradstock, R.A., Gill, A.M., Handmer, J., Hessburg, P.F., Leonard, J., McCaffrey, S., Odion, D.C., Schoennagel, T., and Syphard, A.D. (2014). Learning to coexist with wildfire. *Nature* 525, 58–66.

21. Bond, W. and Zaloumis, N.P. (2016). The deforestation story: testing for anthropogenic origins of Africa's flammable grassy biomes. *Philosophical Transactions of the Royal Society B* 371, 20150170.

22. Davies, G.M., Kettridge, N., Stoof, C.R., Gray, A., Ascoli, D., Fernandes, P.M., Marrs, R., Allen, K.A., Doerr, S.H., Clay, G.D., McMorrow, J., and Vandvik, V. (2016). The role of fire in UK peatland and moorland management: the need for informed, unbiased debate. *Philosophical Transactions of the Royal Society B* 371, 20150342.

23. Johnston, F.H., Henderson, S.B., Chen, Y., Randerson, J.T., Marlier, M., DeFries, R.S., Kinney, P., Bowman, D.M.J.S., and Brauer, M. (2012). Estimated global mortality attributable to smoke from landscape fires. *Environmental Health Perspectives* 120, 695–701; Tse, K., Chen, L., Tse, M., et al. (2015). Effect of catastrophic wildfires on asthmatic outcomes in obese children: breathing fire. *Annals of Allergy, Asthma & Immunology* 114, 308–11.

24. Johnston, F., Melody, S., and Bowman, D.M.J.S. (2016). The pyro-health transition: how fire emissions have influenced human health from the Pleistocene to the Anthropocene. *Philosophical Transactions of the Royal Society B* 371, 20150173.

25. Johnston, F.H., Henderson, S.B., Chen, Y., Randerson, J.T., Marlier, M., DeFries, R.S., Kinney, P., Bowman, D.M.J.S., and Brauer, M. (2012). Estimated global mortality attributable to smoke from landscape fires. *Environmental Health Perspectives* 120, 695–701.

26. Moritz, M.A., Moody, T.J., Krawchuk, M.A., Hughes, M., and Hall, A. (2010). Spatial variation in extreme winds predicts large wildfire locations in chaparral ecosystems. *Geophysical Research Letters* 37, L04801; Peterson, S.H., Moritz, M.A., Morais, M.E., et al. (2011). Modelling long-term fire regimes of southern California shrub-

lands. *International Journal of Wildland Fire* 20, 1–16; Moritz, M.A., Parisien, M.-A., Batllori, E., Krawchuk, M.A., Van Dorn, J., Ganz, D.J., and Hayhoe, K. (2012). Climate change and disruptions to global fire activity. *Ecosphere* 3(6) A49, 1–22; Chornesky, E.A., Ackerly, D.D., Beier, P., et al. (2015). Adapting California's ecosystems to a changing climate. *Bioscience* 65, 247–62; Barros, A.M.G., Pereira, J.M.C., Moritz, M.A., et al. (2013). Spatial characterization of wildfire orientation patterns in California. *Forests* 4, 197–217.

27. Archibald (2016); Bond, W. and Zaloumis, N.P. (2016). The deforestation story: testing for anthropogenic origins of Africa's flammable grassy biomes. *Philosophical Transactions of the Royal Society B* 371, 20150170.

28. There have been significant advances in fire planning over the past few years. See Gazzard, R., McMorrow, J., and Aylen, J. (2016). Emergency planning for wildfire in the United Kingdom: an evolving response from forestry, fire and rescue services. *Philosophical Transactions of the Royal Society B* 371, 20150341.

29. Scott, A.C., Chaloner, W.G., Belcher, C.M., and Roos, C. (2016). The interaction of fire and mankind: introduction. *Philosophical Transactions of the Royal Society B* 371, 20150162.

30. Martin, D.A. (2016). At the nexus of fire, water and society. *Philosophical Transactions of the Royal Society B* 371, 20150172.

31. Moritz, M.A., Batllori, E., Bradstock, R.A., et al. (2014). Learning to coexist with wildfire. *Nature* 515, 58–66. Doerr, S. and Santín, C. (2016). The 'wildfire problem': perceptions and realities in a changing world. *Philosophical Transactions of the Royal Society B.* 371, 20150345; Roos, C.I., Scott, A.C., Belcher, C.M., Chaloner, W.G., Aylen, J., Bliege Bird, R., Coughlan, M.R., Johnson, B.R., Johnston, F.H., McMorrow, J., Steelman, T. and the Fire and Mankind Discussion Group (2016). Contradiction, conflict, and compromise: addressing the many dimensions of human–fire–climate relationships. *Philosophical Transactions of the Royal Society B* 371, 20150469.

32. Scott et al. (2016).

FURTHER READING

There are very few popular books on wildfire and none on fire in the geological past. Further information in subjects relevant to our story written for a student and research audience can be found in the following volumes.

BEERLING, D. (2007). *The Emerald Planet: How Plants Changed Earth's History.* Oxford University Press, Oxford.

BELCHER, C.M. (ED.) (2013). *Fire Phenomena in the Earth System: An Interdisciplinary Approach to Fire Science.* John Wiley and Sons, Chichester.

BERNER, R.A. (2004). *The Phanerozoic Carbon Cycle.* Oxford University Press, Oxford.

BURTON, F.D. (2009). *Fire: The Spark that Ignited Human Evolution.* University of New Mexico Press, Albuquerque.

CERDÀ, A. AND ROBICHAUD, P. (EDS) (2009). *Fire Effects on Soils and Restoration Strategies.* Science Publishers Inc., New Hampshire.

COCHRANE, M.A. (ED.) (2009). *Tropical Fire Ecology: Climate Change, Land Use and Ecosystem Dynamics.* Springer, Berlin.

DUNBAR, R.I.M., ET AL. (2014). *Lucy to Language.* Oxford University Press, Oxford.

KEELEY, J.E., BOND, W.J., BRADSTOCK, R.A., PAUSAS, J.G., AND RUNDEL, P.W. (2012). *Fire in Mediterranean Climate Ecosystems: Ecology, Evolution and Management.* Cambridge University Press, Cambridge.

KENNEDY, R.G. (2006). *Wildfire and Americans. How to Save Lives, Property, and Your Tax Dollars.* Hill and Wang, New York.

PELUSO, B. (2007). *The Charcoal Forest: How Fire Helps Animals and Plants.* Mountain Press Publishing Company, Missoula, MT.

PYNE, S.J. (1982). *Fire in America: A Cultural History of Wildland and Rural Fire.* Princeton University Press, Princeton, NJ.

PYNE, S.J. (1992). *Burning Bush: A Fire History of Australia.* Allen and Unwin, Sydney.

PYNE, S.J. (1997). *Vestal Fire: An Environmental History, Told through Fire, of Europe and of Europe's Encounter with the World.* University of Washington Press, Seattle.

PYNE, S.J. (2001). *Fire: A Brief History*. University of Washington Press, Seattle.

PYNE, S.J. (2002). *Year of the Fires: The Story of the Great Fires of 1910*. Penguin, London.

PYNE, S.J. (2007). *Awful Splendour: A Fire History of Canada*. University of British Columbia Press, Vancouver.

PYNE, S.J. (2012). *Fire: Nature and Culture*. Reaktion Books, London.

SCOTT, A.C., MOORE, J., AND BRAYSHAY, B. (EDS) (2000). Fire and the Palaeoenvironment. *Palaeogeography, Palaeoclimatology, Palaeoecology* 164, 1–412.

SCOTT, A.C. AND DAMBLON, F. (EDS) (2010). Charcoal and its use in palaeoenvironmental analysis. *Palaeogeography, Palaeoclimatology, Palaeoecology* 291, 1–165.

SCOTT, A.C., BOWMAN, D.J.M.S., BOND, W.J., PYNE, S.J., AND ALEXANDER, M. (2014). *Fire on Earth: An Introduction*. John Wiley and Sons, Chichester.

SCOTT, A.C., CHALONER, W.G., BELCHER, C.M., AND ROOS, C. (EDS) (2016). The interaction of fire and mankind. *Philosophical Transactions of the Royal Society B* 371.

WILLIS, K.J. AND McELWAIN, J.C. (2014). *The Evolution of Plants*, 2nd edition. Oxford University Press, Oxford.

WRANGHAM, R.W. (2009). *Catching Fire: How Cooking Made Us Human*. Profile Books, London.

LIST OF FIGURE CREDITS

Carboniferous ecosystems, pp. 91–112, Copyright (1994), figure 6, with permission from Elsevier

38 Redrawn with data from W.G. Chaloner and W. S. Lacey (1973) The distribution of Late Palaeozoic floras. In Hughes, N.F. (ed.), *Organisms and Continents Through Time*. Special Papers in Palaeontology, 12, 241–69

39 Modified from Glasspool, I.J., Scott, A.C., Waltham, D., Pronina, N.V., and Longyi Shao. 2015. The impact of fire on the Late Paleozoic Earth system. *Frontiers in Plant Science* 6, 756

40 Karen Carr, Australian Museum

41 Ian Glasspool and A.C. Scott

42a A.C. Scott

42b A.C. Scott

43 A.C. Scott

44 Reprinted from figure 1 in *Cretaceous Research* 36, Brown, S.A.E., Scott, A.C., Glasspool, I.J., and Collinson, M.E., Cretaceous wildfires and their impact on the Earth system, pp. 162–90, Copyright (2012), with permission from Elsevier

45 Reprinted from figure 1 in *Cretaceous Research* 36, Brown, S.A.E., Scott, A.C., Glasspool, I.J., and Collinson, M.E., Cretaceous wildfires and their impact on the Earth system, pp. 162–90, Copyright (2012), with permission from Elsevier

46 A.C. Scott

47 Photo courtesy of M.E. Collinson

48 A.C. Scott

49 Adapted with new data from Scott, A.C., Bowman, D.J.M.S., Bond, W.J., Pyne, S.J., and Alexander M. 2014. *Fire on Earth: An Introduction*. J. Wiley and Sons

50 Redrawn from figure 2 in Bond, W.J. and Scott, A.C. Fire and the spread of flowering plants in the Cretaceous, *New Phytologist* (Wiley 2010), 188: 1137–50. doi:10.1111/j.1469-8137.2010.03418.x © New Phytologist Trust (2016)

51 From P. Bartlein and J. Marlon

52 Image courtesy of T. Swetnam

53a A.C. Scott

53b A.C. Scott

53c A.C. Scott

53d A.C. Scott

54 www.cartoonstock.com

55 Adapted from the work of J.A.J. Gowlett

56 Adapted from Archibald, S., Staver, A.C., Levin, S.A. 2012. Evolution of human-driven fire regimes in Africa. *Proc. Natl Acad. Sci. USA* 109, 847–52

57 Diagram A.L.R. Westerling

58 Image courtesy of T. Swetnam

59 Image courtesy of Guido van der Werf

60a A.C. Scott

60b A.C. Scott

61 Figure 1 in Bowman, D.J.M.S., Balch, J., Artaxo, P., Bond, W.J., Cochrane, M.A., D'Antonio, C.M., DeFries, R., Johnston, F.H., Keeley, J.E., Krawchuk, M.A., Kull, C.A., Mack, M., Moritz, M.A., Pyne, S.J., Roos, C.I., Scott, A.C., Sodhi, N.S., and Swetnam, T.W. 2011. The human dimension of fire regimes on Earth. *Journal of Biogeography* 38, 2223–36

End Image A.C.Scott

Appendix Ages based on the 2017 International Chronostratigraphic chart produced by the International Commission on Stratigraphy http://www.stratigraphy.org/index.php/ics-chart-timescale

LIST OF PLATE CREDITS

Black and white plates

1 A.C. Scott
2a A.C. Scott
2b A.C. Scott
3a A.C. Scott
3b A.C. Scott
4a A.C. Scott
4b A.C. Scott
4c A.C. Scott
5 A.C. Scott
6a A.C. Scott
6b A.C. Scott
6c A.C. Scott
6d A.C. Scott
6e A.C. Scott
7a A.C. Scott
7b A.C. Scott
7c A.C. Scott
8a A. C. Scott
8b A. C. Scott
9 A. C. Scott
10a A.C. Scott and I.J. Glasspool, Geology, Field Museum of Natural History, Chicago
10b A.C. Scott and I.J. Glasspool, Geology, Field Museum of Natural History, Chicago; Specimen PP55042
10c A.C. Scott and I.J. Glasspool, Geology, Field Museum of Natural History, Chicago
10d A.C. Scott
11 From Collinson, M.E., Steart, D.C., Scott A.C., Glasspool, I.J., and Hooker, J.J. 2007. Episodic fire, runoff and deposition at the Palaeocene–Eocene boundary. *Journal of the Geological Society, London* 164 87–97

Colour plates

1 MODIS Project at NASA Image 1163886
2a NASA
2b Graphic by: Min Minnie Wong from NASA data
3 <https://earthobservatory.nasa.gov/IOTD//view.php?id=5800>
4 © Tom Reichner/shutterstock.com
5 Image courtesy of S. Doerr
6 John McColgan, Bureau of Land Management, Alaska Fire Service. Alaskan Type I Incident Management Team/Wikimedia Commons/Public Domain
7 Image courtesy of S. Doerr
8 Image courtesy of S. Doerr
9 A.C. Scott
10 A.C. Scott
11 A.C. Scott
12 Steve Greb
13 Image courtesy of Douglas Henderson
14 Xinhua/Alamy Stock Photo

PUBLISHER'S ACKNOWLEDGEMENTS

We are grateful for permission to include the following copyright material in this book.

Extract from S. J. Pyne from *Global Biomass Burning: Atmospheric, Climatic, and Biospheric Implications*, edited by Joel S. Levine, (MIT Press, 1991).

Extract republished with permission of University of Washington Press, from Stephen J. Pyne, *World Fire: The Culture of Fire on Earth* (University of Washington Press, 2014); permission conveyed through Copyright Clearance Center, Inc.

Excerpt from 'Fire and Ice' from the book *The Poetry of Robert Frost*, edited by Edward Connery Lathem. Copyright ©1923, 1969 by Henry Holt and Company, copyright ©1951 by Robert Frost, copyright ©1998 by Penguin, Twentieth Century Classics. Used by permission of Henry Holt and Company and Penguin. All rights reserved.

The publisher and author have made every effort to trace and contact all copyright holders before publication. If notified, the publisher will be pleased to rectify any errors or omissions at the earliest opportunity.

INDEX

(Italic numbers refer to Figures)